THERMAL CONDUCTIVITY OF SOLIDS

J.E.Parrott and Audrey D.Stuckes

 Pion Limited, 207 Brondesbury Park, London NW2 5JN

ISBN 0 85086 047 4

Set on IBM 72 Composers by Pion Limited, London.
Printed in Great Britain by J.W.Arrowsmith Limited, Bristol.

THERMAL CONDUCTIVITY OF SOLIDS

J.E.Parrott and Audrey D.Stuckes

Applied physics series

Series editor H.J.Goldsmid

1 **Structure and properties of magnetic materials** D.J.Craik

2 **Solid state plasmas** M.F.Hoyaux

3 **Far-infrared techniques** M.F.Kimmitt

4 **Superconductivity and its applications** J.E.C.Williams

5 **Physics of solid state devices** T.H.Beeforth and H.J.Goldsmid

6 **Hall generators and magnetoresistors** H.H.Wieder

7 **Introduction to noise analysis** R.W.Harris and T.J.Ledwidge

8 **Thermal conductivity of solids** J.E.Parrott and Audrey D.Stuckes

List of symbols

a thermal diffusivity, unit cell dimension
A area
$\mathbf{b}$ Burgers vector with magnitude b
$\mathbf{B}$ magnetic induction
c specific heat
c_p specific heat at constant pressure
c_v specific heat at constant volume
C heat capacity
C_v heat capacity at constant volume
d diameter, lattice spacing
e electron charge
E total or internal energy
$\mathcal{E}$ electron or phonon energy
$\mathcal{E}_F$ Fermi energy
$\mathcal{E}_g$ energy gap
$f(\mathbf{k})$ electron distribution function
f_0 Fermi function
$g(\omega)$ density of states function for normal modes
h Planck's constant
$\hbar$ $= h/2\pi$
H enthalpy
j electric current density
J_n Bessel function of order n
$\mathbf{k}$ wave vector of electrons
k_B Boltzmann's constant
$\mathbf{K}$ reciprocal lattice translation vector
l distance
ℓ mean free path
L length
$\mathcal{L}$ Lorentz number
m^* effective electron or hole mass
M atomic mass
n refractive index
$\mathbf{n}$ unit vector normal to surface
$\tilde{n}$ Planck function
N integer, number of electrons
N_A Avogadro's number
N_d dislocation density
N_s^q phonon distribution function
$N(\mathcal{E})$ density of states function for electrons
p pressure
P number of holes
$\mathbf{q}$ wave vector of phonons
$\mathbf{q}_D$ wave vector at Debye temperature

r	radial coordinate, radius
R_H	Hall coefficient
s	polarisation
t	time
T	absolute temperature
T_c	superconducting transition temperature
u	velocity of molecules, of electrons
u_F	velocity at Fermi energy
U	heat current
v	velocity of phonons
v_s	velocity of sound wave of polarisation s
V	volume
W	thermal resistivity
W_i	intrinsic (ideal) thermal resistivity
W_0	residual thermal resistivity
x	position coordinate
α	optical absorption coefficient
γ	Grüneisen constant
ϵ	emissivity
ζ	Riemann function
θ	temperature, temperature interval
Θ_D	Debye temperature
λ	thermal conductivity
λ_e	electronic thermal conductivity
λ_{ph}	lattice (phonon) thermal conductivity
Λ	Righi–Leduc coefficient
μ	mobility
μ_B	Bohr magneton
ν	oscillator frequency, Poisson ratio
ρ	electrical resistivity
ρ_D	electrical resistivity at Debye temperature
ρ_i	ideal resistivity
ρ_0	residual resistivity
$\hat{\rho}$	density
σ	electrical conductivity
σ_R	Stefan–Boltzmann constant
τ	relaxation time
ϕ	volume fraction, porosity, heat flow density
Φ	heat flow rate
ω	angular frequency
ω_D	Debye frequency

Preface

Modern technology has seen the development of many new materials and the extension of the range of temperatures in common use. As a consequence knowledge of the thermal conductivity of materials from very low temperatures up to the melting or dissociation point has become increasingly important. In some cases a poor thermal conductor is required in order to restrict heat flow, while in others the reverse situation applies and enhancement of heat flow is necessary. Choice of the correct material may be an economic or a safety factor. For example, by building a house of the best thermal insulators, substantial savings can be made in heating costs which may well outweigh the increased cost of building materials. Alternatively, if a suitable thermal environment is to be found for an astronaut in space, safety is of prime importance.

The choice of material may pose problems in meeting the demands of a particular device or process. Suppose, for example, a polymer is being used as an electrically insulating mounting for some electrical device; its temperature may be found to rise too much, due to the impedance presented by the polymer to the flow of heat being generated electrically by the device. The question then arises as to how the heat flow can be enhanced—should a different type of electrical insulator be used; or should an attempt be made to alter the properties of the polymer, by metal loading for example? It is from such simple questions that there has grown a need not only for reliable evaluation of thermal conductivity over a wide temperature range, but also for a basic understanding of the mechanisms involved in heat transfer by conduction.

Anything beyond a superficial glance at the literature will be likely to be a discouraging experience for the uninitiated. There is often an enormous spread of experimental values for the thermal conductivity of even common materials, without it being at all clear whether the divergence is due to differences between samples or to inaccurate measurements. The fact that the basic methods of measuring thermal conductivity are commonly part of the school physics syllabus will have given rise to the expectation that obtaining accurate data is a relatively simple problem, whereas in reality it requires both clear thinking and sophisticated apparatus. On the other hand, the theory of thermal conductivity in solids is unlikely to be dealt with thoroughly in undergraduate courses because of the difficulties involved in a rigorous treatment. These difficulties will be immediately apparent to the nonmathematical reader who tries to get to grips with recent literature dealing with fundamental aspects of the subject. If, however, one is prepared to take fundamentals for granted, one will often find that relatively straightforward physical thinking will suffice in the interpretation of experimental data.

An introductory chapter is followed by a discussion of the basic principles of the methods of measuring thermal conductivity and diffusivity, without experimental detail but with due emphasis on the sources of error and their control. Then the theory is outlined from a physical, rather than a mathematical, point of view, and the ways in which the theory can be applied to the interpretation of data are explained. The last chapter is of particular interest to those concerned with applications of materials where the thermal conductivity is an important parameter. We have shared the responsibility for writing this book as follows: chapters 2, 3, and 4, A.D.S.; chapters 4 and 5, J.E.P.; whilst chapter 1 was written jointly.

The book has been designed to bridge the gap between the knowledge which can reasonably be expected of a final year physics or engineering student and the level implied by most specialised treatises, reviews, and research papers. As indicated above, this means that on the experimental side we have found it necessary to stress the difficulties, whilst on the theoretical side we have tried to reassure the reader that the problems are perhaps less formidable than he might have feared.

Contents

1

Introduction to thermal conductivity

1.1 The significance of heat conduction

Human appreciation of the importance of the conduction of heat begins with the well-known fact that some things, a piece of metal for instance, feel cold to the touch while others, for example a piece of wood, feel warm. The reason for this is that metal conducts heat away from the body faster than wood. All-year-round human occupation of the temperate and colder parts of the Earth's surface has always depended on the ability to control the loss of heat from the body. Amongst the things that feel warm because of their low thermal conductivity are the furs, skins, and woven cloth used by mankind for millenia for protection against external cold. Similar considerations govern the choice of building materials, particularly in more recent times under the impact of technology employing the results of scientific analysis of heat transfer. The reverse problem arises in a spacecraft, where it is necessary to keep the astronauts cool during re-entry into the Earth's atmosphere.

These examples all call for materials of low conductivity, but other situations require a high rate of heat transfer. For example the need to protect certain semiconductor devices from damage due to overheating has led to the semiconductor being mounted on a diamond heat sink, diamond having a higher thermal conductivity around room temperature than any metal. This is an extreme case, but there are many others where efficiency requires the transfer of heat with a minimum temperature difference.

Thus we sometimes need a high, sometimes a low thermal conductivity. Often this will be associated with a requirement of good mechanical strength or high electrical conductivity, and so on. This reveals that the study of thermal conductivity often requires a context of materials science, in the widest sense.

It will be clear that the existence of a body of data on thermal conductivity and related properties will never suffice to meet the problems posed by technology. What is also needed is theoretical understanding which will enable us to predict the thermal conductivity of new materials, and to guide us in our attempts to find materials to cope with new requirements. In turn the testing of theories will challenge the experimenters by requiring new standards of accuracy in measurement.

The interaction of theory and experiment has influenced the layout of this book. It begins with some preliminary questions of definition and so forth and then goes on to describe the methods by which accurate data can be obtained. This is followed by an outline of the theory and of the ways in which it may be applied to experimental results. Finally, the behaviour of everyday practical materials is reviewed and considered.

1.2 Thermal conductivity: Fourier's law

The first clear statement of the proportionality of heat flow and temperature gradient was made in 1822 by Fourier in his *Theorie Analytique de la Chaleur.* It should be realised, however, that this kind of linear law does not apply to other forms of heat transfer such as convection or radiation where, although the heat flow is some function of the temperatures of the two regions involved, this function will generally be by no means simple. In fact in the case of solids a lack of proportionality between apparent heat flow and temperature gradient would often be regarded as evidence that some nonconductive mechanism was at work. This might be due to a deficiency in the experimental arrangement or, very rarely, in the case of some materials transparent in the infrared there might be a genuine component of heat transfer by radiation.

The linear proportionality of heat flow and temperature gradient may be supposed to be observed in a situation where there is a flat slab of material of thickness Δx whose faces are isothermal surfaces but at temperatures differing by an amount ΔT. We suppose that there is some means of measuring the heat flow into and out of these surfaces. If the slab is effectively thermally insulated at the edges and there are no internal sources of heat, such as electric currents or radioactivity, then in a steady state the rate of heat flow Φ into one face equals that out of the other. We then find that for a given slab

$$\Phi \propto \Delta T,$$

and if we take varying thicknesses of slab then

$$\Phi \propto \frac{\Delta T}{\Delta x} .$$

Furthermore, if we now vary the area A of the slab,

$$\Phi \propto A \frac{\Delta T}{\Delta x} ;$$

this relation may then be used to define the thermal conductivity λ thus

$$\Phi = -\lambda A \frac{\Delta T}{\Delta x} .$$

From this line of argument it is possible to generalise to a vector heat current density

$$U = -\lambda \operatorname{grad} T . \qquad (1.1)$$

The minus sign arises from the fact that heat always flows from the hotter to the colder region. Equation (1.1) will be the form of Fourier's law generally used in this book.

There are a number of qualifications which must be made with reference to this line of argument:
(i) Since the thermal conductivity is a function of the temperature, if one works back from equation (1.1) to the original expression in terms of ΔT, this will clearly break down if ΔT becomes large enough to encompass significant changes in λ.
(ii) Some materials are anisotropic with respect to heat conduction and it will be seen that this will mean that the heat flux vector $\boldsymbol{U}$ will not necessarily be parallel to grad T. This requires the generalisation of (1.1) to a form which can be expressed either by

$$\boldsymbol{U} = -\boldsymbol{\lambda}\,\mathrm{grad}\,T, \tag{1.2a}$$

where $\boldsymbol{\lambda}$ is written as a dyadic, or by

$$U_i = -\sum_j \lambda_{ij}\frac{\partial T}{\partial x_j}; \tag{1.2b}$$

we have introduced here a tensor λ_{ij} having nine components, of which no more than six may be different, because where $i \neq j$, $\lambda_{ij} = \lambda_{ji}$. Equation (1.2b) will sometimes be found without the summation being explicitly written in, but then it is understood that repeated suffixes are summed over. Fortunately, for polycrystalline materials and cubic crystals the simple equation (1.1) will suffice.
(iii) Although the vectors $\boldsymbol{U}$ and grad T are defined as though at a point in the solid, there will clearly be difficulties of a conceptual kind if this is regarded too literally, since neither $\boldsymbol{U}$ nor grad T can have any meaning for a single atom in a solid. Theoretical discussion always assumes that these quantities are in fact defined with respect to regions which, although small, contain enough atoms for the fluctuations in $\boldsymbol{U}$ and grad T to be negligible.
(iv) There may be problems relating to measurements on rather small but otherwise completely homogeneous samples where, if the cross-sectional area is decreased, the heat current decreases more than proportionally. This 'size effect', as it is called, really means there is no properly defined thermal conductivity at all, but in practice the concept of a size-dependent 'effective' thermal conductivity is used.

Another question concerns whether there are any subsidiary conditions necessary for the meaningful measurement of thermal conductivity. There appears at present to be only one, and this is that no electric current must be flowing in the material under examination. The reason for this is that, if there is a current, then the Peltier heating where the current enters and leaves the material under investigation may be inadvertently added to the heat carried by conduction. Furthermore, there may be additional interference due to the Thomson effect. Electric currents would not normally be deliberately passed through a specimen during a thermal conductivity experiment without these effects being allowed for, but under some circumstances there might be currents passing due to the Seebeck effect which would pass unnoticed. The most desirable way of proceeding is to ensure open circuit conditions during the measurement of thermal conductivity, unless the passage of a current is essential to the method being employed. As a definition of thermal conductivity we must add to equation (1.1) the condition

$$\boldsymbol{j} = 0, \tag{1.3}$$

where $\boldsymbol{j}$ is the electric current density.

1.3 Conservation of energy and the definition of thermal diffusivity

The linear law relating heat flow and temperature gradient gives only a partial description of the thermal processes involved in solids. In particular it is adequate only for steady state phenomena with no internal sources of heat. To go further requires the use of the principle of conservation of energy, otherwise known as the first law of thermodynamics.

Let us consider a small volume inside the conducting medium (the meaning of 'small' is that discussed in the previous section). Then, if there is no work being done on this volume, the change in its internal energy will be given by the heat transfers across its boundaries. Thus, if ΔE_0 is the internal energy at time $t = 0$, and ΔE_t that at time t, then

$$\Delta E = \Delta E_t - \Delta E_0 = \Delta Q,$$

where ΔQ is the heat entering the small volume. This can be expressed in terms of the time derivative of the internal energy and the heat current integrated over the surface S:

$$\frac{\mathrm{d}(\Delta E)}{\mathrm{d}t} = -\int \boldsymbol{U} \cdot \boldsymbol{n} \, \mathrm{d}S,$$

where $\boldsymbol{n}$ is an outward directed normal to the surface. The term on the left-hand side can be replaced by a volume integral over the internal energy density E, whilst the right-hand side can be replaced by a volume integral, using Gauss's theorem. Then

$$\int \frac{\partial E}{\partial t} \mathrm{d}^3x = -\int \mathrm{div}\, \boldsymbol{U} \, \mathrm{d}^3x,$$

or, since the integration volume is arbitrary,

$$\frac{\partial E}{\partial t} = -\mathrm{div}\, \boldsymbol{U}. \tag{1.4}$$

The changes in internal energy can be expressed in terms of the specific heat c multiplied by the density $\hat{\rho}$:

$$\frac{\partial E}{\partial t} = c\hat{\rho}\frac{\partial T}{\partial t} = -\mathrm{div}\, \boldsymbol{U},$$

or, combining with Fourier's law (equation 1.1),

$$c\hat{\rho}\frac{\partial T}{\partial t} = \text{div}(\lambda \,\text{grad}\, T). \tag{1.5}$$

This equation requires further discussion and elaboration. To begin with, is the specific heat referred to that at constant pressure, c_p, or constant volume, c_v? As the argument was presented above there is no doubt that c_v is appropriate, since work of any kind was excluded, which means no changes in volume. However, this is not the usual experimental situation, since it requires rigid constraints around the conductor to prevent the normal change of volume by thermal expansion. If we use the condition of constant pressure, then the place of the internal energy E must be taken by the enthalpy H, in which case the correct specific heat to employ is c_p. In actual fact a body containing temperature differences normally also contains internal stresses and for that reason c_p is not quite appropriate either. But with a simple one-dimensional temperature gradient c_p is likely to be more nearly correct.

The form of equation (1.5) allows for the possibility of the thermal conductivity varying with position, either owing to the temperature gradient or to actual inhomogeneity of the conductor. However, in most work this effect is neglected and (1.5) is written

$$c\hat{\rho}\frac{\partial T}{\partial t} = \lambda \nabla^2 T,$$

or

$$\frac{\partial T}{\partial t} = a\nabla^2 T, \tag{1.6}$$

where $a = \lambda/c\hat{\rho}$ is called the thermal diffusivity. Equation (1.6) is essential in all discussions of time-varying thermal phenomena in homogeneous media; there are appropriate modifications to allow for anisotropic conductors. In the case of steady temperatures equation (1.6) reduces to Laplace's equation, $\nabla^2 T = 0$.

In deriving equation (1.4) it was assumed that there was no work being done, and it was subsequently shown that the possibility of the performance of work to change the volume affected the proper definition of the specific heat in (1.5). However, there are other examples involving work which can be better regarded as heat generation within the conductor, although this is not a very well-defined concept from the thermodynamic point of view. As an example, if there is an electric current density $\boldsymbol{j}$ and an electrical conductivity σ (assumed scalar), then an external electromotive force is doing a quantity of work j^2/σ in unit volume, which is normally expressed as a heat generation of j^2/σ per unit volume. In this case equation (1.4) becomes

$$\frac{\partial E}{\partial t} + \text{div}\,\boldsymbol{U} = \frac{j^2}{\sigma}. \tag{1.7}$$

For any other process involving work done within the conductor there will be a corresponding term added to equation (1.4) in the same way.

It has been pointed out that an equation such as (1.6) has certain rather implausible consequences. If we consider for example a flat slab and apply at a given instant a supply of heat to one face, then according to (1.6) there is an instantaneous effect at the far face. This of course cannot occur in practice, since no signal can be propagated through the slab at infinite velocity. One way of avoiding this is to modify (1.6) as follows:

$$\nabla^2 T = \frac{1}{a}\frac{\partial T}{\partial t} + \frac{1}{u^2}\frac{\partial^2 T}{\partial t^2}, \qquad (1.8)$$

where u is a quantity having the dimensions of velocity. If u is made equal to the velocity of sound, then the paradox of instantaneous propagation is avoided, but the effect of this term is less than that of the first for times greater than a/u^2. For a good conductor these times are about 10^{-11} s and they are even shorter for poor conductors. For all practical purposes therefore equation (1.6) is quite satisfactory.

1.4 The physical mechanisms of the conduction of heat in solids

In section 1.2 it was shown how it is possible to discuss heat conduction in solids in terms of a single coefficient, the thermal conductivity, and in section 1.3 equations were derived whose solutions describe the temperature distribution in a solid. Subsequently these equations will be applied in analysing the experimental methods used to determine the conductivity. In this section a survey is given of the physical processes involved in heat conduction.

The simplest material to consider is the perfect electrical insulator. Many materials of both technical and scientific interest can be regarded as approximating to this. To understand the transport of heat in such a material one considers the form in which the internal energy exists. This is almost exclusively in the lattice vibrations, as the thermal motion of the atoms or ions is usually called. If a model of the solid is used where the atoms are coupled to their neighbours by forces, which, although of a quantum-mechanical nature, can be treated classically, the resultant expressions for internal energy and specific heat are in good agreement with experiment at both low, intermediate, and high temperatures. Even the Debye model, in which the lattice vibrations are treated as sound waves, gives quite an adequate picture for many purposes.

One of the most important features of models of this kind is that the vibrations are analysed into normal modes obeying harmonic oscillator equations. These harmonic oscillators are found to possess energy only in discrete integer units of $h\nu = \hbar\omega$, where ν is the oscillator frequency, ω $(= 2\pi\nu)$ is the angular frequency, h is Planck's constant, and $\hbar = h/2\pi$. To be precise, the energy of the oscillator must be of the form

$$\mathcal{E}_n = (n + \tfrac{1}{2})\hbar\omega,$$

where n is an integer, and the half gives the inaccessible, but detectable, 'zero point' energy. These quanta $\hbar\omega$ are called 'phonons' in the solid by analogy with the photons of electromagnetic radiation. It is this quantisation which causes the rapid decrease in specific heat at low temperatures. From many points of view these phonons can be regarded as particles and the solid as a gas of such particles. Then heat conduction appears as a diffusion of phonons from a hotter region where they are more numerous to a colder region where they are less so. The alternative classical picture requires consideration of the varying amplitudes of lattice waves in hotter and colder regions and is considerably less vivid.

It may be shown that in an infinite perfect single crystal where the lattice vibrations are strictly harmonic there is no resistance to the flow of phonons. Departure from strict harmonicity gives rise to collisions between phonons and a thermal resistivity, $1/\lambda$, proportional to absolute temperature. This is characteristic of the high-temperature behaviour of insulators.

Phonon scattering due to the presence of impurity atoms and other point defects of the crystal lattice becomes effective at fairly low temperatures. Finally, at very low temperatures the main mechanism of phonon scattering is collision with the surface of the crystal or with grain boundaries inside a polycrystalline insulator. This gives rise to a decrease of thermal conductivity as T^3 at low temperatures. It can be seen that with $\lambda \propto T^{-1}$ at high temperatures, and $\lambda \propto T^3$ at low temperatures, there is a maximum value of λ at some intermediate temperature. This kind of behaviour characterises insulators with fairly good crystal perfection, though not ceramics or glasses.

From the point of view of analysis of the experimental data the easiest materials to understand are the pure metals. It was discovered as early as 1853 that the ratio of thermal to electrical conductivity was very similar for a large number of metals, and it was later shown that this 'Wiedemann–Franz ratio' was proportional to absolute temperature as long as the temperature was not too low. This clearly indicated that the mechanism of heat transport was the motion of the free electrons in the metal. However, this conclusion left two questions unanswered. The first was why this large number of free electrons did not contribute to the specific heat. This problem was solved by the application of quantum mechanics to the statistics of electrons. The second was what had happened to the phonon (lattice vibrational) thermal conductivity, which in many insulators is nearly as large as the thermal conductivity characteristic of pure metals. The answer to this was to be found in the scattering of phonons by electrons, an answer confirmed by the detection of a lattice contribution to the thermal conductivity in some alloys where the electrical conductivity was low, and most unambiguously by experiments on superconductors where, owing to the effective removal of the electrons into a state in which both interaction with phonons and heat transport were impossible, a large lattice thermal conductivity appears.

For most materials it is unnecessary to consider heat conduction mechanisms other than those due to electrons and phonons. As an example where one has to go beyond this, there are certain semiconductors where electrical conduction is due to electrons and positive 'holes' in nearly equal numbers, and here the energy of creating the electron–hole pair contributes to the heat transport.

1.5 General considerations in the measurement of thermal conductivity
The methods of measuring thermal conductivity can be divided into two categories, static and dynamic, depending on whether the temperature distribution within the sample is time dependent. Static measurements involve the use of equation (1.1) and it is necessary to determine the heat current density and the temperature gradient along the normal to the isothermal surface. In contrast to the steady state measurements, dynamic methods involve the complete differential heat flow equation (1.4). In general these methods determine the diffusivity and require measurement of the time for a thermal disturbance to propagate a known distance. The specific heat and density must be known in order to obtain the thermal conductivity, although in some dynamic methods the specific heat can be determined as well as the diffusivity.

Both steady state and dynamic methods require the solution of the appropriate equation for the particular geometry of the sample, heat source, and sink. The simple solutions involve isothermals which are either plane, cylindrical, or spherical surfaces. It is usually an experimental problem to maintain the isothermals of the shape required for a particular mathematical solution, because of heat transfer from the sample to the surrounding medium.

The stationary state condition assists in the achievement of a high degree of precision of measurement, although the total time involved in achieving equilibrium can be a very lengthy process if the conductivity is low. The long time constant also makes steady state methods undesirable at very high temperatures. Dynamic methods, in general, do not give as high a precision as static ones although modern instrumentation is improving enormously the precision attainable with this type of method. There has been a tremendous upsurge of interest in various dynamic techniques in the past few years with the desire to obtain data rapidly, particularly at high temperatures.

The choice of method of measurement depends upon the order of magnitude of the thermal conductivity to be evaluated, on the temperature range, and on the sample size. The latter may depend on the uniformity or macroscopic nature of the material; it may also be restricted by limitations of a manufacturing process.

The thermal conductivity of solids ranges at most over some five orders of magnitude, varying at room temperature from about 4 W cm^{-1} K^{-1} for copper or silver to 10^{-4} W cm^{-1} K^{-1} for microporous materials such as plastic foams. For single-phase solids the spread is only over some three orders—the materials of very low conductivity are those of short range order, such as the polymers and glasses, and multiphase solids with various degrees of porosity. It is not even possible to make the generalisation that metals are better conductors of heat than nonmetals. At room temperature diamond is the best known conductor and, depending upon the quality of the gem, can have a conductivity five times greater than that of copper. Conduction in pyrolytic graphite parallel to the layer planes is of the same order of magnitude as in diamond and is higher than that of copper up to 1200 K. The conduction perpendicular to the layer planes, however, is lower by a factor of about 200.

Although the thermal conductivity of both pure metals and nonmetallic crystals increases with decreasing temperature, close to 0 K many comparatively common electrical insulators conduct heat better than, or certainly as well as, metals. At high temperatures there is a tendency for metals to conduct better than nonmetals; however the range of values steadily diminishes and at 2000 K extends only over two orders of magnitude.

As a result of the comparatively small range of thermal conductivities there is no thermal insulator in the sense of an electrical insulator. Consequently the problem common to all methods of measurement is the attainment of the conditions of heat flow required by the mathematical solutions; moreover the degree of difficulty tends to increase with increasing temperature. As a note of caution it should be remembered that, if these conditions are not met experimentally, then the data acquired are meaningless.

A glance at the wealth of thermal conductivity data published by the Thermophysical Properties Research Centre (TPRC) at Purdue University (Touloukian, 1970) shows a disparity in data probably greater than that of any other physical property. The disagreement is more than a few per cent, may be as high as an order of magnitude, and is in general far larger than the claimed precision of the data. Some differences can be expected as no two samples can be completely identical. However, as it is shown in Chapter 6, for homogeneous materials these differences should be small (except at very low temperatures) and certainly smaller than the literature suggests. Much of the diversity is due to a lack of accuracy in the data, resulting from failure to meet experimentally the required conditions of heat flow, and in this respect TPRC has critically appraised much of the data and published recommended curves. However, this is an age of new materials and, since prediction of thermal conductivity is extremely difficult, it is a property which must be investigated experimentally. In order to make new data more reliable than those of the past, apparatus for the measurement of thermal conductivity or diffusivity should be thoroughly checked for systematic errors. Although these can be difficult to locate, particularly without experience, they can sometimes be found by repeating measurements under different experimental conditions, changing for example sample size or heat flux. The apparatus should finally be checked by measuring one or more materials of known thermal conductivity. There is no single reference material, but the standard or standards chosen to 'check' a particular apparatus should cover the full range of conductivities for which the apparatus is to be used. High-purity copper, Armco iron, various nickel alloys, and particular glasses are commonly used. For work of the highest precision a bank of reference standards is being established at the National Bureau of Standards in Washington from whom details of stock material can be obtained. Over the past few years an international cooperative measuring programme has examined, amongst other thermal properties, the thermal conductivity and diffusivity of a number of metals and nonmetals above room temperature. The report by Fitzer (1973) serves as a useful guide to the choice of reference materials, some of which are readily available.

When reporting thermal conductivity data as much information as possible should be given in order to characterise the material. This should include the source of the material, its chemical analysis, fabrication treatment, density, grain size, crystal structure, and direction of heat flow, together with details of shape, size, and orientation of any pores or additional phases in heterogeneous materials. For electrical conductors the electrical conductivity, Hall coefficient, and thermoelectric power should be specified as functions of temperature, as these are extremely useful aids to material classification.

In the following two chapters the principles involved in various steady state and dynamic methods of measurement are discussed critically without experimental detail.

1.6 A note on units

The situation as regards the units in which thermal conductivity is measured has for some time been very confused. Until recently we have had one system usually employed by engineers in English-speaking countries and two systems employed by scientists and Continental engineers. A further system based on SI Units has now appeared and will have legal force in Britain in the near future.

The first system referred to has as its basis the British thermal unit (Btu), the hour, the foot, and the degree Fahrenheit. Thus the units of thermal conductivity in this system are

$$\text{Btu h}^{-1}\text{ ft}^{-1}\text{ }^{\circ}\text{F}^{-1}.$$

The two systems used by scientists differ in that one uses calories and the other joules (or ergs). In the first of these we have as units of thermal conductivity

$$\text{cal s}^{-1}\text{ cm}^{-1}\text{ K}^{-1},$$

whilst in the second we have

W cm^{-1} K^{-1},

where K stands for the kelvin, defined as the fraction 1/273·16 of the thermodynamic temperature of the triple point of water [this symbol is used both for thermodynamic temperature (in place of °K) and for temperature interval (in place of deg, °C, °K)]. The SI system is based on the second system, but insists on metres rather than centimetres, so the units are

W m^{-1} K^{-1}.

There seems to us no excuse for persisting with the Btu h^{-1} ft^{-1} $°F^{-1}$ system or the calorie-based metric system. At present inconvenience is caused by the existence of great quantities of data expressed in units of these systems, but to continue using them will only increase these difficulties for the future. The two remaining systems only differ by a factor of one hundred and can easily be used in harness. Except in building applications the metre seems rather a large unit to employ. Furthermore most current scientific work seems to use the W cm^{-1} K^{-1} system and for these two reasons we shall use these units. Similarly we shall generally use cm^2 s^{-1} for the units of diffusivity. To assist in the use of the literature a table for converting the systems of units one to another is given in the Appendix.

References

Fitzer, E., 1973, AGARD Advisory Report No.606, March.

Touloukian, Y. S. (Ed.), 1970, *Thermophysical Properties of Matter. The Thermophysical Properties Research Centre Data Series,* volumes 1 and 2 (IFI/Plenum Press, New York).

Static methods of measuring thermal conductivity

2.1 Introduction

In static methods the thermal conductivity is obtained from measurements of a temperature gradient together with the heat flux into or out of the sample in accordance with equation (1.1).

In its simplest form the sample is a cylinder in which the heat flow is parallel to the axis and the isothermals are planes perpendicular to the axis. This is the basis of the so-called linear or axial flow method. Owing to the inevitability of heat losses it is difficult to ensure that the temperature gradient remains normal to the cross-sectional area. Moreover the problems of maintaining uniform isothermals are enhanced as the temperature of measurement is increased.

As an alternative, a sample which surrounds a heat source is used. This could be a hollow sphere with a central spherical heater or a long cylinder with the heater along the axis. In these cases the heat flow is radial and the isothermals are either spheres or infinite cylinders. This is the basis of the radial flow method.

2.2 Linear flow method

2.2.1 General principle

If all the heat supplied by the electric heater at the rate Φ is conducted along a rod of uniform cross-sectional area A with distance L between thermometers, as shown in figure 2.1, then at any point

$$\lambda = -\frac{\Phi}{A}\frac{\mathrm{d}L}{\mathrm{d}T}, \tag{2.1}$$

and the mean conductivity λ between the temperatures T_1 and T_2 is

$$\lambda = -\frac{\Phi \Delta L}{A \Delta T} = -\frac{\Phi L}{A \Delta T} = \frac{\Phi L}{A(T_1 - T_2)}. \tag{2.2}$$

This assumes that the heat losses from the periphery of the sample and along the thermometers are negligible. In order to ensure that the heat flow is linear between the thermometers, they should not be closer to the heater or the heat sink than a length equal to the sample diameter.

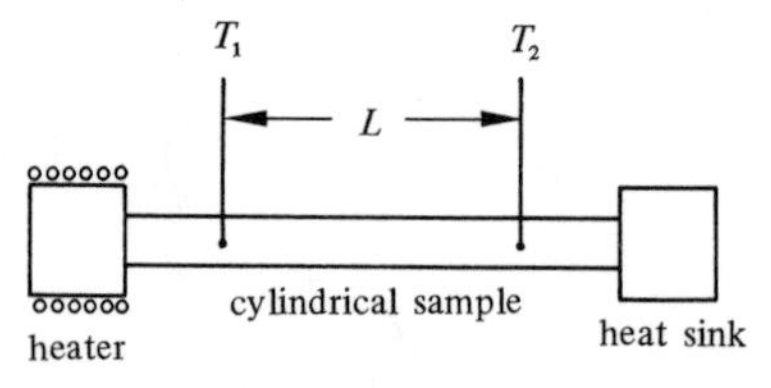

Figure 2.1. Schematic diagram for measurement of thermal conductivity under steady-state linear heat flow.

In general, measurement of the heat flow rate Φ and of the cross-sectional area A presents no problem. However, it is necessary to pay great attention to the measurement of the temperature difference $(T_1 - T_2)$ along the length L. Since thermal conductivity varies with temperature, quite markedly in many cases, the temperature difference should be small so that a meaningful average temperature can be assigned to the measurement. The thermometers, which if possible should fit snugly to a reasonable depth in holes in the sample, should be as small in cross-section as practicable, since ideally the temperature at a point is required and the distance between the two points must be measured precisely. Since the accuracy is dependent upon the measurement of a difference in temperature of only a few degrees, the precision of absolute temperature measurement must be high. Alternatively, if direct measurement of the temperature difference can be made, the absolute value of either T_1 or T_2, from which the average temperature can be assigned, can be determined with less precision. For this reason thermocouples are ideal thermometers, since the temperature difference can be measured by connecting the two thermocouples back to back. It is of course essential to maintain the cold junctions of the thermocouples at the same temperature, preferably by immersing the junctions in a suitable constant-temperature environment of high thermal mass, so that any small fluctuations in temperature have a long time constant compared with the time necessary to make the measurements. By using suitable switches, free from spurious thermoelectric effects, the voltage output from each thermocouple can be measured directly as well as the differential e.m.f. However, at low temperatures the sensitivity of thermocouples becomes comparable to, or less than that of, resistance thermometers, so that it becomes necessary to make precise measurements of each absolute temperature. It must also be remembered that heat flow from, or to, the specimen will be conducted by the thermometer leads, which should be kept as fine as possible.

The choice of specimen geometry is controlled mainly by the conductivity value to be measured, by the sensitivity of the thermometers, and by practicable values of Φ. The ratio of length to area L/A must be large enough to ensure that ΔT is sufficiently large to measure without needing an enormously large heat flux. Lateral heat losses are proportional to the surface area, so that to minimise them a short sample with a large cross-section is required. Measurement of the temperature gradient then becomes difficult since the presence of the thermometers significantly alters the heat flow pattern. This can be overcome by attaching each end of the sample to a good conducting block in which the temperature is measured; the measured temperature gradient is then likely to contain spurious gradients across the contact between the sample and block, the magnitude of which will be more significant the greater the thermal conductivity being measured. From this point of view the arrangement illustrated in figure 2.1 is to be preferred. The wider the temperature range to be covered, the more difficult it becomes to achieve a compromise between the two mutually contradictory requirements. In practice a long bar is used for a good conductor and a thin disc for a bad conductor.

2.2.2 Measurements on good conductors below 300 K

For low-temperature measurements on metals and crystalline solids an arrangement similar to that shown in figure 2.1 can be used. The sample is usually mounted vertically with the heater at the bottom, and the heat sink at the top coupled to a suitable refrigerant. The assembly is placed in an evacuated chamber so that the heat losses from the periphery of the sample are purely radiative. By suitable choice of lead wires to the heater and to the thermometers, and by anchoring them adequately to the refrigerant, the conductive heat losses from the assembly can be made extremely small compared with the heat conducted through the sample. Typical arrangements are those of Rosenberg (1954), White and Woods (1955), and Slack (1957).

Below 100 K the radiative heat loss is usually negligible and equation 2.2 holds. However, from 100–300 K correction must be made for the radiative loss; this is usually found experimentally by changing the L/A ratio of the sample. Alternatively a heated shield must closely surround the sample and heater (with the temperature gradient along the shield approximately matched to that of the sample) as done by Slack (1961). An apparatus used at the National Bureau of Standards, described by Powell *et al.* (1957, 1959), incorporates a much longer sample than usual, with the temperature measured at eight equally spaced positions along its length. In this case a heat shield, with its temperature at two points matched to that at corresponding positions on the sample, is used over the full range from 4 to 300 K.

Various types of thermometers can be used. In earlier measurements gas thermometers were used, particularly at very low temperatures; these are now largely superseded by resistance thermometers or thermocouples. Carbon and semiconducting resistances are in common use as well as such thermocouples as gold–cobalt versus Manganin; and copper, silver, or gold–iron versus Chromel. In general the temperature interval ΔT is measured differentially. Details of suitable thermometers and of cryostat design in general are given in handbooks on cryogenic techniques.

This method is suitable for conductivities above about 10^{-1} W cm^{-1} K^{-1}, and can therefore be used for measurements on all crystalline solids since these are good conductors at low temperatures. The accuracy of the method in the very low temperature region is chiefly limited by measurement of the temperature difference. The measurement of the heat input and sample area presents no problems, but the length between thermometers can be a limiting factor if the sample is very short. In addition, from

100 to 300 K the radiation correction imposes restrictions upon the accuracy. Generally, with care an overall accuracy of 1% can be obtained but precision greater than this is extremely difficult and has rarely been achieved.

2.2.3 Measurements on good conductors above 300 K

As the temperature is increased above 300 K heat losses from the periphery of the sample rapidly increase in magnitude as radiative heat transfer starts to predominate and, even in good conductors, heat losses can become comparable with the heat conducted through the sample. It is therefore necessary to minimise radiation by surrounding the sample with a solid of low conductivity, usually in the form of a powder. Laubitz (1969) has calculated that, even under these conditions, when an unguarded sample is placed in an isothermal furnace at a temperature equal either to that of the cold end of the sample or its mean temperature, large departures from the assumed linear temperature gradient occur when the ratio of sample conductivity to that of the insulating powder is as high as 10^3. It is necessary therefore to place a guard shield between the furnace and the sample as shown in figure 2.2. The shield is heated independently of the sample and ideally the temperature everywhere along the guard should equal that at corresponding points along the sample. A further heated guard is placed over the sample heater to ensure all its heat flows downwards through the sample. Well designed apparatus of this type for use up to 1000°C is described by Ditmar and Ginnings (1957) and by Powell and Tye (1960).

Although guarding the sample may appear to be an easy way to obtain linear heat flow, in practice it is extremely difficult to achieve. From a naive picture of the situation, a radial heat flow outward from the sample

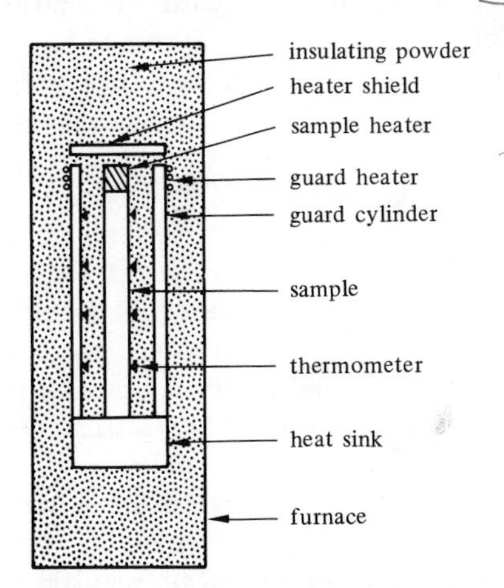

Figure 2.2. Schematic diagram of apparatus with cylindrical heat shield for linear-heat-flow steady-state measurements.

is to be balanced by a radial flow inward from the guard. However, the actual heat flow pattern from the sample heater depends upon its total environment, including the presence of the guard itself when it is close to the sample. If the guard is too large in diameter, appreciable heat loss from the sample will occur by linear heat flow through the insulating medium and it becomes impossible to provide an exactly equal and opposite heat flow pattern. The problem has been analysed mathematically by Laubitz (1969); the numerical solutions which he gives provide useful guidelines to the design of an experimental system and the order of accuracy it should achieve.

Since there are many variables involved it is only possible here to give general guidance on the design of an apparatus. The ratio of sample length to diameter should be about 10, with the outside diameter of the guard about twice that of the sample. The conductivity of the guard should be of the same order as that of the sample. The lower end of the guard should be at the same temperature as the bottom of the sample, while a short heater should be wound around the guard adjacent to the sample heater. The temperature between the guard and sample should be matched at positions fairly close to the top of the sample. If the lower end of the guard and sample are not connected solidly together, a lower heater should be wound around the guard to ensure that the temperatures are also matched here. In general, additional heaters on the guard do not help as they tend to feed too much heat to the sample, thereby distorting the linear heat flow pattern. The guard should be surrounded quite closely by a furnace which has a temperature gradient matched roughly to that of the guard.

The main limiting factor of this method is the thermal conductivity of the insulating medium surrounding the sample and guard. The ratio of the sample conductivity to that of the insulator ideally should be at least 10^3. As the range of conductivities becomes smaller with increasing temperature, a ratio of 10^2 is more realistic. This sets a lower limit of $\sim 10^{-2}$ W cm^{-1} K^{-1} to the value which can be measured by this technique.

In addition to the inherent design error introduced by failure to achieve exact linear heat flow, errors arise in the measurement of temperature gradient and of the heat flowing through the sample. Although measurement of the electrical power input can be made very accurately, it is impossible to ensure that none of this is lost from the upper surface of the heater or along the leads. Also, separation of the thermometers can be extremely difficult to measure in electrically insulating samples as electrical methods for their location cannot be used; at 1000°C a 5% accuracy can be achieved with reasonable care but anything better than this is extremely difficult.

The thermometers for use at room temperature and up to 1700 K are normally thermocouples, as resistance thermometers tend to be either too large or too unstable in this range. Depending upon the temperature range and the environment, these can be nickel–chromium versus Constantan, Chromel versus Alumel, or platinum versus platinum–rhodium (see Caldwell, 1962).

The calibration of all thermocouples can be changed to some degree by mechanical strain and work hardening introduced during the manufacture of the thermocouple junction and its assembly in the apparatus. Consequently all thermocouples should be annealed before use and handled as carefully as possible. Some change in calibration will occur during use, possibly as a result of contamination. Since steady state methods demand measurement of a temperature difference, any inequalities in calibration between two thermocouples monitoring the difference can be taken into account by measuring the thermocouple outputs under isothermal conditions before power is supplied to the heater. This provides a correction to the temperature difference measured when the heater is energised.

2.2.4 Comparative measurements

It is not always possible to obtain samples sufficiently long for the guarded system, particularly when studying single crystals of new materials. Measurements can be made on samples with a smaller length-to-diameter ratio, by making a composite bar consisting of the disc under test sandwiched between two discs of a material of known conductivity, similar in magnitude to that of the test material. The temperature gradient in the unknown disc and in the two standards is then measured. This technique has been used on optical crystals by McCarthy and Ballard (1951), on semiconductors by Stückes and Chasmar (1956), and by Mirkovich (1965) on ceramics.

One difficulty with this method is that when the sample and standards have a length-to-diameter ratio less than one, thermometers inserted in the discs are so close to the interfaces that they could be in a region of nonuniform heat flow. It is possible to place good conductors such as silver, gold, or copper on either side of the specimen and the standards, and to insert the thermometers into these. This has the disadvantage that large errors can arise through thermal contact resistance at each interface. It can be overcome by using between the surfaces a metallic liquid such as an indium–gallium alloy or a ductile solid like indium, as long as it truly wets the surface when molten. But because of diffusion effects the contact resistance increases rapidly above 400°C and limits the usefulness of this technique.

Well specified stainless steels and pure metals make excellent standards in the range 1 – 10 W cm^{-1} K^{-1}. The method is not suitable for higher conductivities, as the temperature difference across the discs becomes to small for accurate measurement with a reasonable heat flux and the contact resistances become more significant. Standards of lower conductivity are not easy to find.

The method is not without criticism, and Laubitz (1969) has briefly analysed some of its defects. It is worth noting that, although the accuracy of a well-designed guarded system cannot be equalled, over a limited temperature range the comparative method is useful; as well as being suitable for measurements on discs a few millimetres thick, it needs little instrumentation. In addition, the presence of a standard on either side of the unknown soon indicates the existence of heat losses, and even in the hands of the uninitiated the method is not likely to give rise to the enormous errors which have been a feature of thermal conductivity data.

2.2.5 Guarded hot plate method

This is the standard method for the measurement of low thermal conductivity materials where it is necessary to use thin samples with a length-to-diameter ratio much less than 1 in order to obtain a reasonably small temperature gradient.

Standard specifications for this type of method are given in BS 874 (1965) and ASTM C177 (1963) and the principle is illustrated in figure 2.3. It consists essentially of two similar samples sandwiched between heat sinks and separated by a transverse heater; this heater has a central section separated from an annular guard by a small gap. The whole assembly is surrounded by material of low conductivity. The two sections of the heater are supplied from independent stabilised power supplies which are adjusted until the temperature of the guard and the central section is the same, thereby eliminating radial heat losses from the central heater. The thermal conductivity is then found from equation (2.2) where Φ is the power supplied to the central heater, A is the cross-sectional area of this heater plus half that of the gap, L is the total thickness of the two samples and $(T_1 - T_2)$ is the temperature difference across each sample.

For measurements at room temperature and below, liquids at an appropriate temperature are circulated through the heat sinks (figure 2.3a).

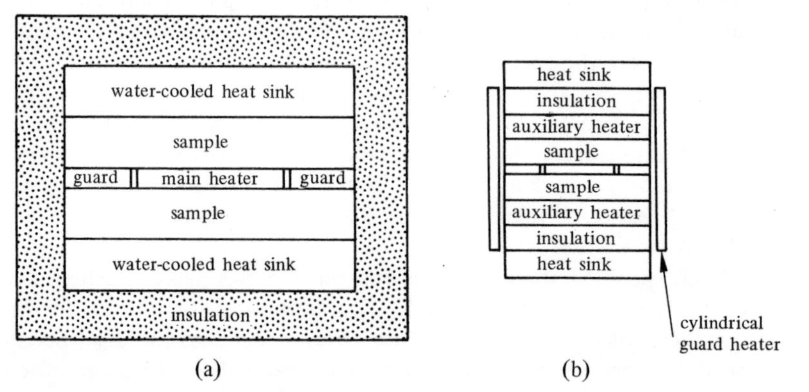

Figure 2.3. Schematic diagram of guarded hot plate apparatus for measurements at room temperature and below (a) and at higher temperatures (b).

At higher temperatures heated plates are inserted between the heat sinks and the samples (figure 2.3b). In order to maintain axial flow over the measured area throughout the samples, these must not be too thick and the guard must be sufficiently wide. The British Standard requires a ratio of diameter of the central heater to sample thickness of about 3, together with a limiting value of 1 for the ratio of guard ring width to sample thickness. Analysis of edge losses by Woodside (1957) indicates that this value is too small and that the ASTM requirement of a limiting ratio of 1·5 is more realistic if an accuracy of 1% is to be achieved. Woodside also shows that it is necessary to control the edge temperature of the sample. For measurements above room temperature the gap between the central and guard sections of the heater should be filled with loose insulation and an additional cylindrical guard heater must be provided in order to control the edge temperature conditions. Tye (1970) has shown that this must be closely controlled at the mean sample temperature in all measurements above 500°C.

It should also be remembered that the thermal conductivity of good insulating materials often depends markedly upon such parameters as shape and size of pores, grains, and fibres, so that materials must be well specified if results are to be compared. Furthermore the results obtained can be affected by the gaseous environment, the presence of moisture, and the orientation of the sample, as well as by the temperature conditions within the apparatus.

2.3 Radial flow method

In general radial flow methods are applied to measurements above room temperature since the longitudinal method presents few problems at low temperatures. In this type of method the flow of heat is radial and, in accordance with equation (1.1), grad T represents the temperature gradient along the normal to the isothermal surface. The most common geometry used consists of a right circular cylindrical sample with the heat source along the axis, as shown in figure 2.4. If its length is sufficient to be assumed to be infinite, then the solution to the Fourier equation is

$$\frac{\Phi}{L} = \frac{2\pi\lambda(T_1 - T_2)}{\ln(r_2/r_1)} \tag{2.3}$$

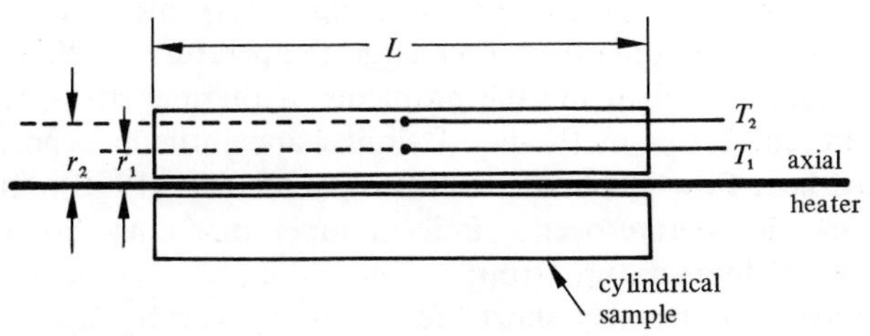

Figure 2.4. Schematic diagram of radial heat flow apparatus.

where Φ/L is the heat flow per unit time per unit sample length, and T_1 and T_2 are the temperatures at radii r_1 and r_2 respectively.

For equation (2.3) to apply the ratio of sample length to its diameter L/d must be large. However, the diameter of the sample cannot be too small or the radial temperature gradient becomes too difficult to measure with any accuracy, so that in general large samples are required.

In order to satisfy the infinite geometry requirements, Kingery (1954) in measurements on alumina used an L/d ratio of 12 and the sample was 18 in long and of 1·5 in diameter. One of the smallest samples investigated by this technique, as reported by Slack and Glassbrenner (1960), was a cylinder of germanium 6 cm long and 1·5 cm in diameter; but as a result of problems associated with the measurement of the temperature gradient, they ultimately increased the sample diameter to 2·5 cm and maintained an L/d ratio of 4. Details of the apparatus are given by Glassbrenner (1965) and show that sample fabrication and assembly are not easy.

To overcome some of the problems associated with the manufacture of cylindrical samples of a suitable size and with the assembly of the apparatus, a specimen composed of a stack of discs is commonly used. One such apparatus was designed by R. W. Powell (1939) at the National Physical Laboratory. Although using a sample 6·3 cm in diameter and 22·5 cm long, it also employed end guard heaters to remove any axial temperature gradient. The technique was unique in that the axial gradient was probed by a single movable thermocouple. The results which he obtained on Armco iron are still quoted as standard and have resulted in the use of Armco iron as a 'standard' of thermal conductivity. A basically similar, well designed apparatus using a sample of similar dimensions is described by Godfrey *et al.* (1965). Although this apparatus was structurally sound for use up to 1400°C, thermocouple instability limited its operation to 1100°C.

Measurements up to 2600°C have been achieved by this technique with an apparatus designed by Rasor and McClelland (1960). It incorporated a comparatively short sample, 7·6 cm long and 5·1 cm in diameter, placed in a graphite helical heater with end guard heaters. The heat flow was radially inward and was measured by a water calorimeter on the specimen axis. The usual outward radial flow was not used, as a heater material in the form of a thin rod, capable of operating at very high temperatures, was not available. Consequently a very large temperature gradient was present in the sample, enhancing the problems of thermal stress normally encountered in radial flow methods. A similar apparatus incorporating radial outward heat flow has been described by Moeller (1970). It is hoped that this will operate over a similar temperature range but with only a 50 K radial temperature drop; the graphite rod axial heater is expected to have an extremely short life. In both cases temperature measurements above 1000°C are made pyrometrically by sighting on small-diameter holes within the sample.

At first sight, radial systems such as a sphere with an internal spherical heat source, from which all the heat generated must pass through the sample, would appear to have many advantages. In practice, however, fabrication is difficult, particularly of a uniformly heated source. In addition, measurement of the temperature gradient is uncertain as a result of the curvature of the isotherms and heat leakage along the thermocouple wires, which must lie on a thermal gradient. There are not many references to such systems. Kingery (1954) used a spherical arrangement for measurements on an alumina sphere. McQuarrie (1954) investigated several oxides using a prolate spheroid geometry. This had the advantage that the isotherms crossing the minor axis were relatively flat. His apparatus was also unique in that the heat was generated inductively in a solid molybdenum core.

2.4 Special methods for electrical conductors

The measurement of thermal conductivity by methods in which the sample is heated by the passage of an electric current is growing in popularity. The method is usually attributed to Kohlrausch (1900); the thermal conductivity is evaluated from measurements of the temperature and voltage gradients along the sample. The basic theory is too long to be included here, but Flynn (1969) gives an admirable account of the mathematical solutions to the heat flow equation under various boundary conditions corresponding to the periphery of the sample, either thermally insulated or losing heat by conduction and radiation. In addition Flynn discusses critically the various experimental arrangements which have been used, and compares the results obtained by these methods with other data.

The main problem which has been associated with the direct heating method is the complex mathematics involved. The practice has been to simplify this by imposing restrictive conditions which are difficult to meet experimentally. There is now a trend to rely on computers to handle the sophisticated mathematics which arise from practical rather than idealised systems. An overall view of direct heating methods has been given by Taylor (1972). These methods, as stressed by Powell and Taylor (1970), are undoubtedly important in view of the large number of properties which can be investigated simultaneously.

References

Caldwell, F. R., 1962, *Temperature-Its Measurement and Control in Science and Industry*, volume 3, Part 2 (Reinhold, New York), pp.81-134.
Ditmar, D. A., Ginnings, D. C., 1957, *J. Res. Nat. Bur. Stand.*, **59**, 93.
Flynn, D. R., 1969, in *Thermal Conductivity*, Ed. R. P. Tye, volume 1 (Academic Press, New York), pp.241-300.
Glassbrenner, C. J., 1965, *Rev. Sci. Instrum.*, **36**, 984.

Godfrey, T. G., Fulkerson, W., Kollie, T. G., Moore, J. P., McElroy, D. L., 1965, *J. Am. Ceram. Soc.,* **48**, 297.
Kingery, W. D., 1954, *J. Am. Ceram. Soc.,* **37**, 88.
Kohlrausch, F., 1900, *Ann. Phys.,* **1**, 132.
Laubitz, M. J., 1969, in *Thermal Conductivity,* Ed. R. P. Tye, volume 1 (Academic Press, New York), pp.111-185.
McCarthy, K. A., Ballard, S. S., 1951, *J. Opt. Soc. Am.,* **41**, 1062.
McQuarrie, M., 1954, *J. Am. Ceram. Soc.,* **37**, 84.
Mirkovich, V. V., 1965, *J. Am. Ceram. Soc.,* **48**, 387.
Moeller, C. E., 1970, in *Proceedings of European Conference on Thermophysical Properties of Solids at High Temperatures*, Baden-Baden, 1968 (Deutsche Keramische Gesellschaft), p.125.
Powell, R. L., Roder, H. M., Hall, W. J., 1959, *Phys. Rev.,* **115**, 314.
Powell, R. L., Rogers, W. M., Coffin, D. O., 1957, *J. Res. Nat. Bur. Stand.,* **59**, 349.
Powell, R. W., 1939, *Proc. Phys. Soc.,* **51**, 407.
Powell, R. W., Taylor, R. E., 1970, *Rev. Int. Hautes Tempér. et Réfract.,* **7**, 298.
Powell, R. W., Tye, R. P., 1960, *The Engineer,* **209**, 729.
Rasor, N. S., McClelland, J. D., 1960, *Rev. Sci. Instrum.,* **31**, 595.
Rosenberg, H. M., 1954, *Proc. Phys. Soc.,* **A67**, 837.
Slack, G. A., 1957, *Phys. Rev.,* **105**, 832.
Slack, G. A., 1961, *Phys. Rev.,* **122**, 1451.
Slack, G. A., Glassbrenner, C. J., 1960, *Phys. Rev.,* **120**, 782.
Stuckes, A. D., Chasmar, R. P., 1956, *Report of Meeting on Semiconductors* (Physical Society, London), p.119.
Taylor R. E., 1972, *High Temperatures - High Pressures,* **4**, 523-531.
Tye, R. P., 1970, *Rev. Int. Hautes Tempér. et Réfract.,* **7**, 308.
White, G. K., Woods, S. B., 1955, *Can. J. Phys.,* **33**, 58.
Woodside, W., 1957, ASTM-STP 217 (American Society for Testing and Materials, Philadelphia), pp.49-62.

3

Dynamic methods of measuring thermal conductivity

3.1 Introduction

In dynamic or nonsteady state methods of measurement, the temperature distribution throughout the sample varies with time; consequently the complete differential equation of heat flow (equation 1.4) is involved and the diffusivity is measured. In general, therefore, knowledge of the specific heat per unit volume is necessary if the thermal conductivity is to be obtained. Although this can be a disadvantage, specific heat is less sensitive to impurities and to structure than thermal conductivity; it is also comparatively independent of temperature above the Debye temperature. Consequently available specific heat data can often be used, so that it is not always necessary to determine experimentally specific heat as well as diffusivity. In accurate work, however, particularly on impure materials or composites, it is essential that the specific heat is also measured.

As the measurement time is short, heat losses have a smaller influence on the measurement than in steady state methods—moreover they can be included in the differential heat flow equation or its boundary conditions. As a result the heat loss coefficient appears in the solution of the equation and can be eliminated by making additional measurements under different experimental conditions. This is particularly desirable at high temperatures, and, generally speaking, dynamic techniques are most useful at room temperature and above, where heat losses make the steady state methods more difficult. This, coupled with the shorter overall time required for measurements, has led to the development of a large number of dynamic techniques in recent years. Many of these are reviewed by Danielson and Sidles (1969).

Dynamic methods can be divided broadly into two categories, periodic or transitory, depending upon whether the thermal energy is supplied to the sample with a modulation of fixed period or as a single addition or subtraction. As a consequence the temperature changes in the sample are either periodic or transitory. In general, measurements of the power input are not required in evaluations of thermal diffusivity nor are absolute measurements of temperature, but relative changes in magnitude of temperature as a function of time and position must be recorded. The temperature detectors therefore do not need accurate calibration; all that is required is a sensitive detector whose response is linear over changes of a few degrees. Consequently radiation-sensitive devices such as photocells and photomultipliers can often be used.

3.2 Periodic methods

3.2.1 General principle

The basic principle of this method, devised originally by Ångström (1863), is that if one end of a sample is heated periodically, then the temperature along the sample also varies with the same period but with diminishing amplitude. Moreover, as the temperature wave travels along the sample with finite velocity there is a varying phase relationship. Measurement of the amplitude decrement and either the phase difference or velocity enables the diffusivity to be determined.

As with static methods it is necessary to use a solution of equation (1.4) which is appropriate to a particular experimental arrangement and to the boundary conditions. Some of these are described in the next sections.

3.2.2 Long thin cylindrical samples

If the end of a long thin rod is heated periodically, the one-dimensional form of the heat flow equation can be used and is given by

$$a\frac{\partial^2\theta}{\partial x^2} = \frac{\partial\theta}{\partial t} + \mathcal{H}, \tag{3.1}$$

where $\mathcal{H}$ represents the heat lost from the sample walls and θ is the temperature difference between the sample and the surrounding medium. In order to simplify the equation, it is assumed that Newton's law of cooling applies and that the heat loss varies linearly with the temperature difference θ. Then

$$a\frac{\partial^2\theta}{\partial x^2} = \frac{\partial\theta}{\partial t} + h\theta, \tag{3.2}$$

where the effective heat loss coefficient h must include losses due to radiation as well as convection and conduction. Consequently θ must be kept small so that even at high temperatures, when radiation predominates, linear heat loss is a reasonable assumption.

The solution of this equation is discussed by, amongst others, Ångström (1863), King (1915), Starr (1937), and Sidles and Danielson (1954). Assuming the hot end of the rod has a sinusoidal component of temperature of angular frequency ω superimposed upon a steady temperature θ_0 such that

$$\theta(0, t) = \theta_0 + \theta_1 \cos\omega t, \tag{3.3}$$

and assuming that the cold end of the rod is at the temperature of the surroundings so that

$$\theta(\infty, t) = 0,$$

then

$$\theta(x, t) = \theta_0 \exp(-m_0 x) + \theta_1 \exp(-m_1 x)\cos(\omega t - \varphi x + \psi), \tag{3.4}$$

where θ_0, θ_1, and ψ are constant; m_0 represents the attenuation per unit length of the steady temperature component:

$$m_0 = \left(\frac{h}{a}\right)^{1/2}, \tag{3.5}$$

and thus depends on the diffusivity of the sample and the heat loss coefficient; m_1 represents the attenuation per unit length of the periodic component of temperature:

$$m_1 = \left[\frac{(h^2+\omega^2)^{1/2}+h}{2a}\right]^{1/2}; \tag{3.6}$$

and φ is the phase shift per unit length:

$$\varphi = \left[\frac{(h^2+\omega^2)^{1/2}-h}{2a}\right]^{1/2}. \tag{3.7}$$

m_1 and φ depend on the frequency as well as the diffusivity and heat loss coefficient, both increasing with increasing frequency. Note that with increasing heat losses the attenuation increases and the phase shift decreases, while their product remains constant.

The velocity of propagation v of the temperature wave is given by

$$v = \frac{\omega}{\varphi}. \tag{3.8}$$

The amplitude decrement κ between positions x_1 and x_2 is

$$\kappa = \frac{\exp(-m_1x_1)}{\exp(-m_1x_2)} = \exp[m_1(x_2-x_1)] = \exp(m_1l), \tag{3.9}$$

where $l = x_2-x_1$.

The velocity and the amplitude decrement can be measured, but both depend not only on the diffusivity but also on the heat loss parameter. Since h is not generally known and may change with time, it is essential to determine it experimentally. This can be accomplished by measuring either the velocity or the amplitude decrement at two different frequencies. If v_1 and v_2 are the velocities at angular frequencies ω_1 and ω_2, and the amplitude decrements are κ_1 and κ_2, then by solving the pairs of simultaneous equations it can be shown that

$$\begin{aligned} a &= \frac{v_1v_2}{2}\left(\frac{v_1^2-v_2^2}{\omega_1^2v_2^2-\omega_2^2v_1^2}\right)^{1/2}, \\ &= \frac{l^2\omega^2}{2\ln\kappa_1\ln\kappa_2}\left[\frac{(\omega_2/\omega_1)^2-(\ln\kappa_2/\ln\kappa_1)^2}{(\ln\kappa_2/\ln\kappa_1)^2-1}\right]^{1/2}. \end{aligned} \tag{3.10}$$

Early workers such as King (1915) and Starr (1937) favoured the use of two frequencies, King measuring the velocities and Starr the amplitude decrements.

The use of two frequencies necessarily increases the time of measurement and, at high temperatures particularly, this can be a disadvantage, as not only may the surface of the sample change with time thereby altering h, but also the bulk properties of the sample may be changed. Consequently it is often preferable to use one frequency only, which effectively means the measurement of both the attenuation and the phase shift of the temperature wave. By eliminating h from equations (3.6) and (3.7) and substituting for the attenuation in terms of the amplitude decrement, equation (3.9), it can be shown that

$$a = \frac{\omega}{2\varphi m_1} = \frac{\omega l}{2\varphi \ln \kappa} \,. \tag{3.11}$$

Alternatively, using equation (3.10) this can be written

$$a = \frac{vl}{2 \ln \kappa} \,. \tag{3.12}$$

The latter equation appears to have the advantage that the frequency need not be known explicitly, although it must be constant, and that only the velocity and amplitude decrement of the temperature wave have to be measured.

However, in general, measurement of velocity is more difficult than that of frequency, and it is more usual to use equation (3.11) and to measure both the amplitude decrement and the phase shift. One way of doing this is to feed the periodic components of the temperature at two points distance l apart into the X and Y plates of an X–Y recorder. The resultant trace is an ellipse with its axes inclined to the X, Y coordinate directions. The amplitude decrement is the ratio of the maximum displacements parallel to the X and Y axes, X_{max} and Y_{max} whilst the relative phase is determined from the intercepts on the X and Y axes, X_c and Y_c since

$$X_c = X_{max} \sin\delta, \qquad Y_c = Y_{max} \sin\delta, \tag{3.13}$$

whence

$$a = \frac{\omega l^2}{\delta \ln(X_{max}/Y_{max})} \,. \tag{3.14}$$

A more accurate way to obtain the phase angle δ is to apply a variable-phase reference signal to, say, the X plate of the recorder and one of the temperature signals to the Y plate. The reference phase is then adjusted until a straight-line Lissajous figure results. The second signal is then treated in the same way, and the difference in the reference signals is equal to the required phase angle. The amplitude decrement is now obtained from the ratio of the maximum deflections in the Y direction.

An apparatus for measurements on metals which has proved very successful up to 1300 K is described by Sidles and Danielson (1954, 1960); their technique uses frequencies of the order of 0·008 Hz corresponding to a period of about 2 min. This necessitates the use of samples up to 30 cm long and 0·3 cm in diameter for very good conductors and somewhat shorter for poor conductors. Samples of this size are not always practicable, particularly when considering nonmetals. Abeles *et al.*

(1960) used a similar technique on semiconductors but by employing higher frequencies the temperature wave was attenuated more rapidly so that shorter samples could be used. They give a more detailed analysis of the heat flow equation subject to radiative heat loss only. Although the solution is similar to that given in equations (3.4) to (3.9), it includes a geometrical function enabling samples other than circular in cross-section to be used. To make the effects of changes in surface characteristics small, the heat losses must be low, that is the effective heat loss coefficient h must be small compared with ω, as can be seen from equations (3.6) and (3.7). Abeles *et al.* point out that this gives rise to the constraint

$$\omega r \gg \frac{8\sigma_R \epsilon T_0^3}{C}$$

where r is the radius of a sample of circular cross-section, σ_R is the Stefan–Boltzmann constant, ϵ is the total emissivity, C is the heat capacity per unit volume and T_0 (K) is the mean temperature of the sample. They also show (by allowing the thermal wave to be reflected from the cold end of the sample) that for the sample to approximate to semi-infinite the constraint for its length L is given by

$$\exp(-2m_1 L) \ll L.$$

Abeles *et al.* used modulation frequencies from 0·02 to 0·1 Hz and samples from 5 to 1 cm in length, covering the conductivity range 4 W cm^{-1} K^{-1} to 5 x 10^{-3} W cm^{-1} K^{-1}.

In general, in order to meet the restrictions imposed by the mathematical solution of the heat flow equation, it is necessary that the rise in temperature of the sample above the surrounding temperature be small to preserve the validity of the linear heat loss assumption. Also, any changes in ambient temperature must be small compared with the smallest amplitude being recorded during the time measurements are being made. This implies a very stringent control on the surrounding temperature, since amplitudes of only 0·25°C may be measured. But, as long as the surroundings have a reasonably large thermal capacity, the changes in ambient temperature will be slow, and such variations on the temperature signal can be filtered out by means of a filter tuned to the frequency of the heater supply. This is easier to accomplish the higher the modulating frequency.

The solution also implies a sinusoidal modulation, and if a pure waveform is not achieved then a Fourier analysis should be made. However, the higher harmonics are attenuated more rapidly than the fundamental so that, as long as measurements are not made too close to the heated end of the sample, the temperature wave may well approximate to sinusoidal. Bosanquet and Aris (1954) have given a method of calculation which can be applied to a square wave modulation without recourse to Fourier analysis.

Periodic heating can be provided in various ways. A variable transformer or circuit resistance may be adjusted by an appropriately shaped cam driven by a synchronous motor: alternatively a light source of reasonable intensity, such as a projector lamp, can be focused onto one end of the sample and interrupted by a synchronously driven chopper. A novel form of generator described by Green and Cowles (1960) uses the Peltier effect. A semiconductor p–n thermojunction is heated and cooled by periodic reversals of the current through the junction. This has the advantage over other methods that the time average of the heat supplied during a cycle can be made equal to zero, so that the mean temperature of the sample remains at ambient. This method is limited to the temperature range over which the particular thermojunction operates efficiently; this is about 200°C for Bi_2Te_3 type semiconductors, 400°C for PbTe, and 600°C for Ge–Si alloys.

Thermocouples are normally employed to measure the temperature, and, since the difference between the sample and ambient is required, the cold junctions of the thermocouples can be placed in a suitable heat sink in the vicinity of the sample. Since the periodic variation only is required for the evaluation of diffusivity, it is necessary to back off the e.m.f. corresponding to the steady difference in temperature between the sample and the reference cold junction with the aid of stabilised voltage supplies. Another thermocouple must be used to monitor the temperature of the heat sink so as to provide the sample temperature. The thermocouple junctions must be in very good contact with the sample and have a small thermal capacity so that thermal lag is not introduced by the measuring device. If possible, very small thermocouple junctions should be welded to the sample. The choice of thermocouple wire may depend not only on its sensitivity but also on the compatibility of the thermocouple material and the sample.

3.2.3 Thin plate samples

It is not always possible to provide a sample in the form of a long rod. Moreover, as the temperature of measurement is increased it is difficult to provide a well-controlled uniform temperature over a large volume. Nevertheless samples in the form of thin plates can be used to obtain the diffusivity, as long as higher frequencies are used. This method is particularly useful at temperatures over 1000°C.

Cowan (1961) treats the problem theoretically, assuming linear heat flow, and neglecting heat losses from the sides of the sample, but allowing for different heat losses from the front and back faces. He considers the amplitude and phase of the temperature at both faces; each depends upon the diffusivity and angular frequency as well as the two heat loss parameters. Owing to the use of higher frequencies, small temperature amplitudes must be expected, and detector sensitivity must ultimately be a limiting factor in their measurement. Consequently the measurement of phase is more realistic than measurement of amplitude.

According to Cowan, the phase difference δ_0 between the temperature fluctuations at the front face of the sample and the modulating beam is given by

$$\tan\delta_0 = \frac{2B^3Q_1 + 2B^2\alpha Q_2/(1+R) + \beta BQ_3/R}{2\alpha B^2Q_0 + \beta(1+2R)BQ_1/R + \alpha\beta Q_2/(1+R) + 2B^3Q_3} \tag{3.15}$$

and that between the back face of the sample and modulating beam δ_L is given by

$$\tan\delta_L = \frac{\beta(\tan B - \tanh B) + 2\alpha B\tan B\tanh B + 2B^2(\tan B + \tanh B)}{\beta(\tan B + \tanh B) + 2\alpha B - 2B^2(\tan B - \tanh B)}. \tag{3.16}$$

In these expressions

$$B = L\left(\frac{\omega}{2a}\right)^{1/2},$$

$$Q_0 = \cosh^2 B\cos^2 B + \sinh^2 B\sin^2 B,$$

$$Q_1 = \cosh B\sinh B + \cos B\sin B,$$

$$Q_2 = \cosh^2 B\sin^2 B + \sinh^2 B\cos^2 B,$$

$$Q_3 = \cosh B\sinh B - \cos B\sin B,$$

α and β are related to the heat loss coefficients for the front and back faces, and R is the ratio of heat lost from the back face to that lost from the front face.

For radiative heat loss only and with both faces 'seeing' surroundings at the same temperature, R will not vary significantly from unity. Cowan shows that under these conditions

$$\alpha \approx \frac{8\sigma_R\epsilon T^3L}{\lambda}, \qquad \beta \approx \frac{\alpha^2}{4},$$

where T (K) is the average sample temperature, and L the thickness of the sample.

The front face of the sample can be heated by an electron beam which raises the temperature of the sample above ambient; a modulation of the beam then produces temperature fluctuations of a few degrees in the bombarded face of the sample. Alternatively the front face of the sample can be illuminated by a beam of light. In this case, the temperature of the sample is usually raised above ambient by heating it in a furnace, and when it is in equilibrium a modulated beam of light produces small temperature fluctuations in the irradiated face.

As with measurements on long bar samples, the heat loss parameters must be known if the diffusivity is to be obtained. This can be done by measurements at different frequencies. Cowan has shown that these parameters can also be determined from measurements of the temperature change at the back face of the sample as a function of time, following a sudden irradiation of the front face with a steady source of radiation. This could be preferable to the use of more than one frequency.

Detailed examination of the expressions for the phase differences δ_0 and δ_L shows that over a wide range of frequencies both δ_L and $(\delta_0 - \delta_L)$, the phase difference between the temperatures of the two faces, are comparatively insensitive to the heat loss parameters, while δ_0 is sensitive to them. Thus the accuracy to which these parameters must be known is reduced if measurement of δ_L or $(\delta_0 - \delta_L)$ is made. The choice between these depends on the method chosen for the periodic heating. If the surface is thermally irradiated, reflection of the incident radiation can lead to erroneous measurements of the temperature of the front face. In this case it is preferable to measure the temperature of the back face only and to compare its phase with the modulating signal. This method has been used by Chafik *et al.* (1969) in measurements up to 2100 K in which the source was a xenon lamp. Where an electron beam is used this problem does not arise and either δ_0 or $(\delta_0 - \delta_L)$ may be measured.

In order to use the electron beam technique, the sample must be electrically conducting and experiments must be carried out in a high vacuum. This does not restrict the method to metals, as many insulators at room temperature become sufficiently conducting above 1000°C, although in this case a furnace must be provided to preheat the sample until it is in a sufficiently conducting state. For very high temperature measurements, the vacuum environment could be restrictive, as the vapour pressure of solids becomes so high that a high gaseous pressure is frequently required to prevent evaporation of the sample.

Wheeler (1965) describes an apparatus using the electron beam technique which he has used successfully with a modulating frequency of 0·48 Hz to study several metals up to a temperature of 3000 K. He uses a novel and comparatively simple iterative process to obtain the heat loss parameter.

In general, to satisfy the use of the one-dimensional heat flow equation, the sample should be large in cross-section compared with its thickness and its front surface should be uniformly heated. A phase difference which can be measured with accuracy is required, and this can be obtained by suitable choice of sample thickness and modulating frequency. In view of the higher frequencies (0·1 to 1000 Hz) and higher temperatures normally encountered in measurements on thin samples compared with the long bar techniques, temperature is usually measured by radiation methods involving photocells. Even below 1500°C where reliable thermocouples can be used, their response is usually too slow.

Although the expressions explicitly given by Cowan are for sinusoidal modulation, the same expressions can usually be applied where square-wave modulation is used, for example where the front surface of the sample is irradiated by a chopped light beam. A Fourier analysis can be applied but the harmonics are attenuated so rapidly that this is not usually necessary.

It is possible that as the melting point of solids is approached, vacancy formation occurs and this would be a time-dependent process. A study of the diffusivity by a modulated beam technique over a range of frequencies would be a useful tool for the study of such a process.

3.3 Transitory methods

3.3.1 General principle

All nonequilibrium methods which do not employ periodic heating come under the general heading of transitory methods. In these, the sample is initially in equilibrium with uniform temperature surroundings, then part of it is subjected to a change in thermal flux. The diffusivity is evaluated from changes in temperature which occur in measured time intervals at one or more points within the sample. These methods can be broadly divided into two categories depending on whether the change in thermal flux is maintained for a reasonable time or is supplied as a single pulse; they will be discussed separately.

3.3.2 Step function heating (long bars, thin plates, and cylinders)

In these methods the thermal flux generated at a surface within the sample by a suitable heater, or at the surface by irradiation for example, is suddenly changed to a different, constant level. Solutions of the differential equation of heat flow subject to various boundary conditions are given by Carslaw and Jaeger (1959); these equate the temperature distribution throughout the sample as a function of time, the diffusivity of the sample, and its dimensions. Apparatus must be designed to meet the boundary conditions imposed by the chosen mathematical solution, and then the experimental temperature measurements must be correlated with the theoretical relationship in order to obtain the diffusivity.

The sample can be in the form of a long bar to one end of which a constant thermal flux can be applied. For the simple one-dimensional equation of heat flow to apply, no heat losses must occur, that is the sample must be guarded; this is simpler to accomplish than in steady state methods. A guard cylinder is made of the same material as the sample and, if a single uniform plate heater is placed across the ends of the sample and guard, the temperature distributions along each can be matched. (The use of a single heater in this way is not possible in steady state methods, as normally a measurement of the heat flux through the sample is required.) The temperature is then measured as a function of time at various points along the sample. To cover a range of temperature, the sample is placed in a furnace and brought into equilibrium with its surroundings before the plate heater is energised. Changes in temperature of only a few degrees would normally be measured. Kennedy *et al.* (1962) used this technique for measurements on metals up to 1300 K. A digital computer matched the measured values of temperature at three positions with those computed from the heat flow equation for different values of diffusivity.

As an alternative approach with this long-bar type arrangement, temperature measurements can be taken over a much longer time interval. After a sufficiently long time from the initiation of the thermal flux, such that for a sample of length L

$$\exp\left(-\frac{\pi a t}{L^2}\right) \rightarrow 0,$$

then, as shown by Carslaw and Jaeger (1959), the temperature at any point along the sample increases linearly with time. After time t, if T_1 and T_2 are the temperatures at two points x_1 and x_2, and Ξ is the slope of the temperature-time graph, then

$$a = \frac{\Xi(x_1^2 - x_2^2)}{2(T_1 - T_0)} . \tag{3.17}$$

This method was used by Butler and Inn (1959), who irradiated one end of a guarded sample with the aid of a carbon arc source. This raised continuously the mean sample temperature, and heat loss corrections were necessary.

For all measurements at high temperatures or on low-diffusivity materials guarding the sample to prevent lateral heat flow becomes impracticable. If the sample is in the shape of a flat slab which is short compared with its width, it can be treated as an infinite plate and the one-dimensional heat flow equation is applicable. The temperature rise at some distance x from the heater plane is then given by

$$T = \frac{2\Phi_0}{\lambda} (at)^{\frac{1}{2}} \operatorname{ierfc} \frac{x}{2(at)^{\frac{1}{2}}} , \tag{3.18}$$

where Φ_0 is the heat flux, and ierfc is the integrated error function.

Since the thermal conductivity is not usually known and the heat flux may be difficult to measure, it is usual to measure the ratio of the temperatures at two positions or at two time intervals. In either case it is necessary to compute the value of the temperature ratio as a function of $x/(at)^{\frac{1}{2}}$ from equation (3.18). The experimental value of the ratio is then matched with the computed value so that the appropriate value of $x/(at)^{\frac{1}{2}}$ can be found. Since x and t can be measured, as well as the temperature ratio, the diffusivity can be evaluated. In order to satisfy the given mathematical solution the heater should be a plane, so that a thin foil is required. This means that a low-voltage high-current supply is necessary to obtain sufficient power. In general this technique is particularly useful for low-conductivity materials. It requires a symmetrical assembly, and a foil heater is placed between two thin identical samples which are sandwiched between two heat sinks. A width-to-thickness ratio of ~4 satisfies the condition of linear heat flow.

Harmathy (1964), using this method to study concrete and building materials up to 1000°C, measured at various times the temperature in a plane a known distance l from the heater. From equation (3.18) it follows that

$$\frac{T(2t)}{T(t)} = \frac{2^{1/2}\,\mathrm{ierfc}[l/2(2at)^{1/2}]}{\mathrm{ierfc}[l/2(at)^{1/2}]}, \tag{3.19}$$

where t is measured from the time a constant energy is supplied to the heater. He also analysed the conditions necessary to obtain true linear flow over the measured area.

Glasses and ceramics have been studied up to 1000°C by Plummer *et al.* (1962), who chose to measure the ratio between the temperature in the foil heater T_h and that in the foil heat sinks T_s. Under these conditions, it follows from equation (3.18) that

$$\frac{T_s}{T_h} = \pi^{1/2}\,\mathrm{ierfc}\frac{L}{2(at)^{1/2}}, \tag{3.20}$$

where L is the sample thickness.

The use of step function heating is not confined to linear heat flow but can be extended to conditions of radial flow. As with steady state methods, the only practicable way to achieve this is with a sufficiently long cylindrical sample, heated uniformly either along its axis or over its periphery. The equation of heat flow is then given by

$$\frac{\partial^2 T}{\partial r^2} + \frac{1}{r}\frac{\partial T}{\partial r} = \frac{1}{a}\frac{\partial T}{\partial t}, \tag{3.21}$$

and solutions appropriate to the chosen boundary conditions yield the temperature as a function of the radial position r. Techniques similar to those described for linear flow in long bar samples can therefore be used by substituting the appropriate solution of equation (3.21). For example the temperature can be measured as a function of time at three different radial positions after a constant energy has been supplied to the axial heater. A computer can then be programmed to match the experimental data with those computed from equation (3.21) for various values of diffusivity. This method has been tested up to 1000°C by Carter *et al.* (1965) who used a cylindrical sample composed of a number of discs of Armco iron; the results agreed well with those obtained by the Ångström method of Sidles and Danielson (1954).

The usefulness of this technique for measurements at temperatures where radiative transfer is large has been demonstrated by Cape *et al.* (1963). The peripheries of cylindrical samples of various refractory solids were heated radiatively from a heater coupled inductively to a radiofrequency coil. The diffusivity was evaluated from continuous recordings of the temperature–time response at two radial positions, as the mean sample temperature was increased from 1200 to 1700°C.

One of the problems in using radial heat flow is accurate measurement of the radial positions of the temperature sensors; the precise position of a thermocouple junction within a hole in the interior of the sample is difficult to locate, and if pyrometric methods are used sight holes of finite diameter are required. Samples of fairly large cross-sectional area are therefore necessary; in order to meet the radial flow conditions they must also have a length-to-diameter ratio certainly greater than unity, which makes them fairly massive. In addition sample preparation is not easy.

3.3.3 Determination of thermal conductivity

If heat input is measured as well as temperature changes, step function methods can be used to determine thermal conductivity directly. The basic principle of this method consists of sandwiching the sample between a heat source maintained at a constant temperature and a heat sink of known thermal capacity. Heat flows through the sample, changing the temperature of the heat sink; if there are no heat losses, the heat flowing through the sample is equal to the heat gained by the sink, so that the thermal conductivity can be found if the temperature of the source and the change of temperature of the sink in a known time are measured. As with steady state methods, heat losses—not only from the sample walls but also from the heat sink—are one of the main problems. Heat loss from the sink is particularly serious as the sink serves as the heat meter for the heat flowing through the sample. There are various ways in which the heat losses can be reduced. For example a small sample in the form of a thin disc can be sandwiched between a large copper block of known thermal capacity and a smaller copper block. A copper guard cylinder with a closed end is then placed so that it surrounds completely the smaller copper block and the periphery of the sample as shown in figure 3.1. Its temperature is matched to that of the smaller block. With the system initially in equilibrium, the massive copper block is then placed in a cold liquid such as iced water or liquid nitrogen, and the temperature of the smaller copper block is measured as a function of time as heat flows from it through the sample and into the cold reservoir, until ultimately its temperature approaches that of the large copper block. This technique was used by Ioffe and Ioffe (1958) for measurements in the vicinity of room temperature. Detailed theoretical analyses of the method by Kaganov (1958) and Swann (1959) have shown that there is only a limited time interval over which measurements can be made and a reasonable accuracy achieved. Measurements must not be made too soon after placing the massive copper block in the cooling liquid and they should not be continued once the temperature of the two copper blocks has almost equalised. Under these conditions the thermal conductivity is given by

$$\lambda = \frac{L(C_1 + \frac{1}{3}C_2)}{A(T_1 - T_0)} \frac{dT}{dt} , \tag{3.22}$$

where L is the sample thickness, A is the cross-sectional area of the sample, C_1 and C_2 are the heat capacities of the smaller copper block and the sample respectively, T_1 is the temperature of the small copper block, and T_0 is the steady temperature of the large copper block. The heat capacity of the sample is required as some of the heat flowing into the reservoir comes from the sample, but as it is usually very much smaller than that of the copper block it is only necessary to know it approximately. Allowance must be made for the thermal resistance of the contacts between the sample and the two copper blocks; if measurements are not carried out *in vacuo* a small correction must be made for the thermal conductance of the surroundings.

Ioffe and Ioffe (1958) used this method extensively to study semiconductors, and it is most accurate for conductivities between 0·01 and 0·1 W cm^{-1} K^{-1}. Although the temperature range is restricted, mainly to the vicinity of room temperature, the apparatus is simple to build and to operate. Consequently this method can provide fairly rapid measurements of thermal conductivity, rather than diffusivity, on comparatively small samples.

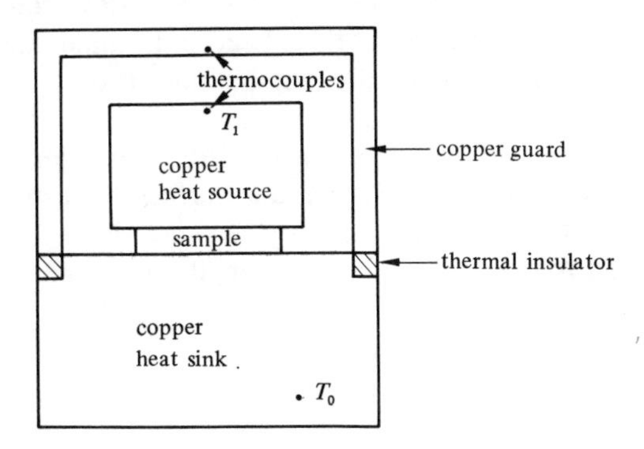

Figure 3.1. Schematic diagram of apparatus for the measurement of thermal conductivity using step function heating.

3.3.4 Pulse and flash heating

In these methods the thermal flux at a surface of the sample is maintained only for a finite time. The temperature at any point within the sample increases to a maximum some time after the removal of the heat source, and then falls, the shape of the time–temperature curve being dependent upon the diffusivity of the sample and the heat losses.

If the end of a rod of cross-sectional area A is subjected to a short heating pulse at $t = 0$ and if there are no lateral heat losses, it can be shown (Carslaw and Jaeger, 1959, p.259) that the temperature at time t, at some point x along the bar, is

$$T(x, t) = \frac{Q}{2A\rho c_p(\pi a t)^{1/2}} \exp\left(-\frac{x^2}{4at}\right) , \tag{3.23}$$

where Q is the energy contained in the pulse. The form of the temperature–time curve is shown in figure 3.2. The maximum occurs at time $t_{max} = x^2/2a$. This is difficult to locate with any accuracy, but below the maximum there are two times, t_1 and t_2, at which the temperature has the same value; from equation (3.23) the diffusivity is then given by

$$a = \frac{x^2}{2\ln(t_2/t_1)}\left(\frac{1}{t_1}-\frac{1}{t_2}\right) . \tag{3.24}$$

It is therefore necessary only to measure the temperature as a function of time at some known distance along the bar.

This method has been tested on metals at room temperature by Woisard (1961) who approximated to axial flow conditions by using two rods, each $\frac{3}{16}$ in in diameter and 25 cm long, on either side of a silicon carbide disc heater. Pulses from 1 to 5 ms were produced by discharging a condenser through the heater. Thermocouples monitored the temperature at various positions, their output being fed to a chart recorder moving at 4 in min^{-1}. The method is neat and the diffusivity is easy to calculate from the data. However, it has the disadvantage that axial flow conditions are required and heat losses are ignored, so that, bearing in mind that a long sample is used, the method is not really suitable above room temperature.

An alternative approach devised by Parker *et al.* (1961), and generally called the 'flash' method, consists of irradiating one surface for a time which is small compared with the transit time of the pulse through the sample. A sample in the form of a thin disc is normally used and the

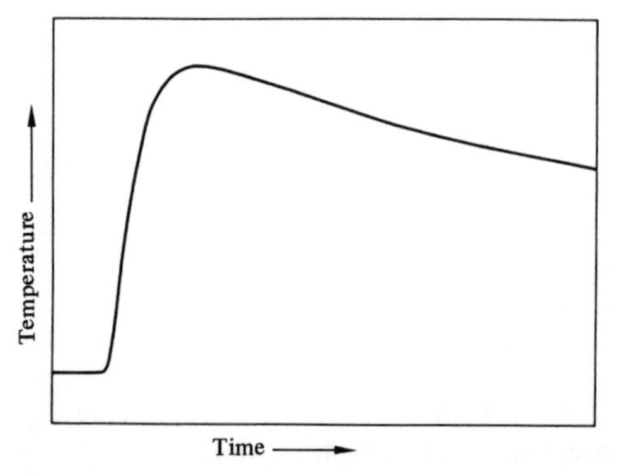

Figure 3.2. Typical temperature–time curve at some distance from the end in a cylindrical sample following the application of a short heat pulse at one end.

temperature is monitored on the rear face of the sample. Heat losses are minimised by making the measurements in a time so short that little cooling can occur. As shown by Parker, the temperature history of the rear face of a sample length L is given by

$$T(L,t) = \frac{\phi}{\rho c_p L}\left[1 + 2\sum_{n=1}^{\infty}(-1)^n \exp\left(-\frac{n^2\pi^2 at}{L^2}\right)\right], \tag{3.25}$$

where ϕ is the heat energy per unit area falling on the front surface. This can be expressed in dimensionless form as a fraction of the maximum temperature T_{max} of the rear surface:

$$\frac{T(L,t)}{T_{max}(L,t)} = 1 + 2\sum_{n=1}^{\infty}(-1)^n \exp\left(-\frac{n^2\pi^2 at}{L^2}\right), \tag{3.26}$$

and this is shown in figure 3.3.

As long as heat losses are negligible, the temperature rises to a maximum value which remains constant. When $T/T_{max} = 0{\cdot}5$, $\pi^2 at/L^2 = 1{\cdot}37$, so that the diffusivity is given by

$$a = \frac{1{\cdot}37L^2}{\pi^2 t_{1/2}}, \tag{3.27}$$

where $t_{1/2}$ is the time taken for the rear surface to reach half its maximum temperature. Alternatively, the tangent to the temperature–time curve at the point of maximum gradient can be extrapolated to the time axis where the intercept, as shown in figure 3.3, occurs at 0·48 from which

$$a = \frac{0{\cdot}48L^2}{\pi^2 t_x}, \tag{3.28}$$

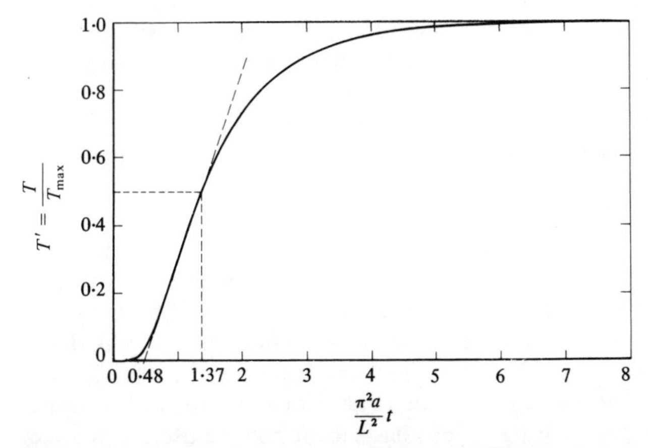

Figure 3.3. The rear face temperature history following flash irradiation of the front face and assuming no heat losses, as given by equation (3.26).

where t_x is the time axis intercept of the measured temperature–time curve. Note that the evaluation of diffusivity does not involve an absolute measurement of temperature but depends only on the ratio of two temperature measurements.

If the energy absorbed by the front face is known then the product of the density and specific heat of the sample is given by

$$\rho c_p = \frac{\phi}{L T_{\max}}, \tag{3.29}$$

so that the thermal conductivity can be evaluated.

If heat losses are not negligible, the maximum temperature reached by the sample will not remain constant but will fall with increasing time. In addition the temperature maximum will be reduced and it will take a longer time to reach it. Under these circumstances it is probably preferable to evaluate the diffusivity by measurement of t_x rather than $t_{1/2}$. Since a small error in slope can lead to considerable errors in t_x and hence a, effort should be made to reduce the heat losses by reduction of sample thickness, and thereby achieving a plateau in the temperature–time curve. It must be remembered that equations (3.25) to (3.28) apply only if the time of irradiation is short compared to the transit time of the pulse through the sample. When this condition is not satisfied the temperature rise is retarded. Cape and Lehman (1963) have shown that as long as the transit time of the pulse t_c defined by $t_c = L^2/\pi^2 a$, is at least ten times greater than the duration τ of the pulse, then equations (3.27) and (3.28) apply. If this condition is not attained then the numerical factor in equation (3.27) increases as τ/t_c is increased, and its value is given graphically by Cape and Lehman.

If heat losses remain important after the maximum possible reduction in sample thickness has been made, the problem is not easily overcome. Using a radiative heat loss parameter, which depends upon the temperature, the emissivity, and the thermal conductivity of the sample, Cape and Lehman give the numerical factor in equation (3.27) as a function of this parameter, and show that it decreases as the heat losses increase. However, as accurate determination of this parameter cannot usually be made, this method can only be used for rough measurements.

An alternative approach is to make temperature measurements for a sufficient time beyond the peak so that the ratio of the temperatures at $5t_{1/2}$ and $t_{1/2}$, or at $10t_{1/2}$ and $t_{1/2}$ can be made as proposed by Cowan (1963). This does not involve a heat loss parameter, but it does have a weak dependence upon the ratio of the heat lost from the back and front surfaces of the sample; this can usually be made equal to unity. Instead of using the temperature ratios, Watt (1966) has shown how to use a digital computing technique to correlate the experimental temperature–time characteristic with a theoretical curve which includes radiative heat losses and a finite pulse time, thereby introducing sophisticated treatment of the experimental data in order to obtain diffusivity.

The flash method is basically simple and offers a means of making rapid determinations of thermal diffusivity on small specimens. The block diagram of figure 3.4 shows schematically the essential requirements. A pulsed source of radiating power is focused onto the front face of a sample within a furnace, while a detector records the changes in temperature at the rear face and the time at which they occur.

The heat pulse can be obtained from a flash tube or a laser. The latter has the advantage that the source can be located some distance from the specimen and still provide sufficient irradiation at its front surface to produce measurable changes at its rear surface. This enables the specimen to be placed well within the uniform temperature zone of the furnace. One problem associated with the use of a laser is that the energy density across the sample face is not uniform. This can lead to quite large errors when the rear face temperature is sensed over a small area, as with a thermocouple, as shown by Beedham and Dalrymple (1970). The error is reduced by using a detector which senses the average rear face temperature. Shaw and Goldsmith (1966) used a laser energy of some 10 J in 2 ms for measurements up to 1000°C on metals and silicon carbide; more power would be required for lower diffusivities and higher temperatures where heat losses are higher. A flash tube dissipating at least 1500 J is necessary, and this must be placed fairly close to the surface (Taylor, 1965).

Up to 1000°C thermocouples can be used as the temperature sensors, but in order to minimise thermal lag in the junction open-ended thermocouples are normally used; fine bent thermocouple wires are

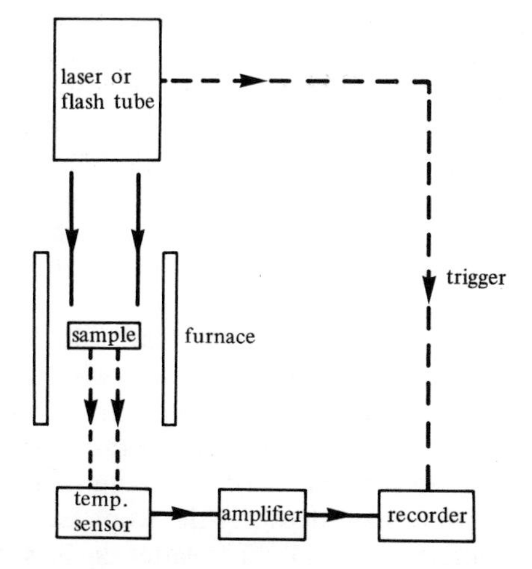

Figure 3.4. Schematic diagram of apparatus for the measurement of thermal diffusivity by the flash method.

pressed against the surface which, if the sample is nonconducting, is coated with a metallic layer or is rubbed heavily with a soft pencil. Alternatively, radiation detectors can be used; these have the advantage that as no contact with the sample is required they introduce no time delays. Below 1000°C infrared-sensitive detectors such as lead sulphide or indium antimonide photocells are necessary, the latter requiring cooling in liquid nitrogen. Above 1000°C it is possible to use photomultipliers sensitive to the visible region of the spectrum. Absolute calibration of the detectors is not required for the evaluation of diffusivity, since it is only necessary to measure the change in temperature of the back face. Since this is small, a few degrees at most, the detectors can be considered to be linear over this range, although it is essential that any associated amplifiers do not introduce nonlinearity. A second detector, either a thermocouple or pyrometer, depending upon the range, monitors the mean sample temperature.

Depending upon the particular sample and its thickness, measurements of temperature are required over a period from about 100 ms to a few seconds. The response of the detector can therefore be displayed either oscillographically and photographed, or on a high-speed ultraviolet recorder, the detecting system being triggered by the flash.

Although the surroundings of the sample must remain at constant temperature during the measurements, the period involved is so short that this is not a stringent requirement. This, coupled with the small sample size and the consequent small uniform temperature zone, offers advantages to this method in very high temperature measurements and an apparatus capable of operating from 1000 to 3000°C is described by Weilbacher and Craeynest (1970).

In general the flash method requires fairly simple specimens, furnaces, and associated controls, but it does need sophisticated electronics associated with the measuring techniques and sophisticated treatment of the results if heat losses cannot be neglected. A most comprehensive and critical review of the pulse method has been given by Righini and Cezairliyan (1973). This contains not only a useful summary of materials which have been studied by this method together with the variations in technique used, but it also lists all the items to which attention must be given if the accuracy is to be assessed.

Undoubtedly it has been proved that, up to 1000°C, the method can produce results, at least on metals, which agree well with those obtained by other methods. A word of caution is necessary in considering the method for either poor conductors or higher temperatures. The solutions of the heat flow equation used in this method assume that the energy is absorbed instantaneously in a small depth at the front face and that the temperature rise above ambient is small. The latter enables the heat loss parameter to be assumed to be constant and the heat loss to depend linearly on the temperature difference. The depth in which a given power is absorbed will decrease as the thermal conductivity decreases, so that the volume of the sample in which the initial dissipation of energy occurs becomes smaller. Furthermore as the thermal conductivity goes down, the sample can maintain a larger gradient between its front and back faces so that more power is necessary to produce a measurable rise on the back face. Consequently a situation could occur where the temperature rise of the front face is so large that the validity of the use of the mathematical solution [equation (3.23)] is in question. If this is taken to its logical conclusion there is the possibility of vaporising the front face of the sample. Even if this extreme situation does not occur, the problem of assigning an effective temperature to a particular measurement must remain. However, a cooperative measuring programme reported by Weisenburger (1973) has shown results for the diffusivity of polycrystalline graphite obtained by the pulse method to agree with those from periodic methods to within $\pm 3\%$ up to 2600 K.

An alternative approach at very high temperatures is a negative pulse technique proposed by Lincoln *et al.* (1974). In this a thin slab is heated by a steady energy flux on one or both faces. The flux to one face is momentarily interrupted and radiative heat losses cause a negative temperature pulse to develop through the sample; the diffusivity can then be evaluated from the temperature change of the other face. As it requires a reasonable heat loss, there is a minimum temperature below which the sensitivity of the method is too low to be practicable—this depends upon the dimensions, diffusivity and emissivity of the sample. The method has been used by Lincoln *et al.* to measure the diffusivity of graphite above 2400 K.

3.4 Thermal conductivity probe

Measurements of the thermal conductivity of poor conductors, such as rocks and building materials, are usually carried out by the guarded hot plate technique. However, a serious drawback to the steady state method is the long time required to attain equilibrium, particularly if any moisture is present in the sample. In addition, where damp solids are under investigation, nonuniform moisture distribution can arise when a temperature gradient is maintained for a long time, as discussed in section 6.3.2. Therefore a dynamic method of measurement is often desirable, and the transient line-heat-source probe, the so-called thermal conductivity probe, is commonly used for this type of material and for powders.

The method employs a line heat source, normally an electrically-heated fine wire, and a temperature sensor placed alongside it and embedded in the material under test. Carslaw and Jaeger (1959) show that when a constant power Φ per unit length is supplied to the heater, the temperature rise θ after time t at a point at a distance r from the heat source in an infinite extent of solid is given by

$$\theta(r,t) = -\frac{\Phi}{4\pi\lambda}\,\mathrm{I}\left(-\frac{r^2}{4at}\right), \tag{3.30}$$

where I is an exponential integral, which for small values of $r^2/4at$ may be approximated to

$$\theta = \frac{\Phi}{4\pi\lambda}\left[\ln\left(\frac{4at}{r^2}\right) - y\right], \tag{3.31}$$

where y is Euler's constant.

In this way the thermal conductivity can be obtained from the slope of a graph of the temperature rise against $\ln t$. Alternatively, the temperature rise $\Delta\theta$ between times t_1 and t_2

$$\Delta\theta = \frac{\Phi}{4\pi\lambda}\ln\frac{t_2}{t_1} \tag{3.32}$$

may be used.

In order to justify the approximations, the distance r must be small and t large. Theoretically, an infinite line heater and an infinite sample are necessary, so that the probe must have a reasonable length-to-diameter ratio and be small compared to the sample.

Temperature rises of a few degrees in about 20 min can be obtained with well designed probes, making them extremely useful tools for the study of moist and powdered solids. However, as always with this type of method, it is essential to ensure that the behaviour of the probe does fit the mathematical solution and boundary conditions of the heat flow equation; it is usually necessary therefore to calibrate the probe. The design factors are discussed at length by Pratt (1969).

References

Abeles, B., Cody, G. D., Beers, D. S., 1960, *J. Appl. Phys.*, **31**, 1585.

Ångström, A. J., 1863, *Phil. Mag.*, **25**, 130.

Beedham, K., Dalrymple, I. P., 1970, *Rev. Int. Hautes Tempér. et Réfract.*, **7**, 278.

Bosanquet, C. H., Aris, R., 1954, *Brit. J. Appl. Phys.*, **5**, 252.

Butler, C. P., Inn, E. C. Y., 1959, *Thermodynamic and Transport Properties of Gases, Liquids and Solids* (ASME, New York).

Cape, J. A., Lehman, G. W., 1963, *J. Appl. Phys.*, **34**, 1909.

Cape, J. A., Lehman, G. W., Nakata, M. M., 1963, *J. Appl. Phys.*, **34**, 3550.

Carslaw, H. S., Jaeger, J. C., 1959, *Conduction of Heat in Solids*, 2nd edition (Oxford University Press, London).

Carter, R. L., Maycock, P. D., Klein, A. H., Danielson, G. C., 1965, *J. Appl. Phys.*, **36**, 2333.

Chafik, E., Mayer, R., Pruschek, R., 1969, *High Temperatures - High Pressures*, **1**, 21.

Cowan, R. D., 1961, *J. Appl. Phys.*, **32**, 1363.

Cowan, R. D., 1963, *J. Appl. Phys.*, **34**, 926.

Danielson, G. C., Sidles, P. H., 1969, in *Thermal Conductivity*, volume 2, Ed. R. P. Tye (Academic Press, New York), pp.241-300.

Green, A., Cowles, L. E. J., 1960, *J. Scient. Instrum.*, **37**, 349.

Harmathy, T. Z., 1964, *J. Appl. Phys.,* **35**, 1190.
Ioffe, A. V., Ioffe, A. F., 1958, *Soviet Phys. Tech. Phys.,* **3**, 2163.
Kaganov, M. A., 1958, *Soviet Phys. Tech. Phys.,* **3**, 2169.
Kennedy, W. L., Sidles, P. H., Danielson, G. C., 1962, *Adv. Energy Conv.,* **2**, 53.
King, R. W., 1915, *Phys. Rev.,* **6**, 437.
Lincoln, R. C., Donaldson, A. B., Heckman, R. C., 1974, *J. Appl. Phys.,* **45**, 2321.
Parker, W. J., Jenkins, R. J., Butler, C. P., Abbott, G. L., 1961, *J. Appl. Phys.,* **32**, 1679.
Plummer, W. A., Campbell, D. E., Comstock, A. A., 1962, *J. Am. Ceram. Soc.,* **45**, 310.
Pratt, A. W., 1969, in *Thermal Conductivity,* volume 1, Ed. R. P. Tye (Academic Press, New York), pp.376-388.
Righini, F., Cezairliyan, A., 1973, *High Temperatures - High Pressures,* **5**, 481.
Shaw, D., Goldsmith, L. A., 1966, *J. Scient. Instrum.,* **43**, 594.
Sidles, P. H., Danielson, G. C., 1954, *J. Appl. Phys.,* **25**, 58.
Sidles, P. H., Danielson, G. C., 1960, in *Thermoelectricity,* Ed. P. H. Egli (John Wiley, New York), pp.270-287.
Starr, C., 1937, *Rev. Scient. Instrum.,* **8**, 61.
Swann, W. F. G., 1959, *J. Franklin Inst.,* **267**, 363.
Taylor, R., 1965, *Brit. J. Appl. Phys.,* **16**, 509.
Watt, D. A., 1966, *Brit. J. Appl. Phys.,* **17**, 231.
Weilbacher, J., Craeynest, J., 1970, *Rev. Int. Hautes Tempér et Réfract.,* **7**, 268.
Weisenburger, S., 1973, *High Temperatures - High Pressures,* **5**, 475.
Wheeler, M. J., 1965, *Brit. J. Appl. Phys.,* **16**, 365.
Woisard, E. L., 1961, *J. Appl. Phys.,* **32**, 40.

4

Theory of thermal conductivity

4.1 Introduction

The role of the theory of thermal conductivity is the same as that of physical theory in general, that is to explain what is observed and to enable predictions to be made. It is the latter function which is primarily of interest to the applied physicist who will commonly be faced with the problem of too much or too little heat conduction in some practical situation. The contents of this chapter and the next will show that the explanatory power of thermal conductivity theory is very great, and that, although its predictive capacity is rather less, it is great enough to be of considerable value in many areas. However, it must be said that the theory shows up best in what are really rather simplified situations.

The basic problem in thermal conductivity theory is twofold. Firstly, one wants to know what it is that carries the heat. A number of possible mechanisms suggest themselves of which far the most important are those by the electrons in metals and the thermal vibrations of the atoms in insulators. Secondly, it is necessary to find out what limits the heat carrying capacity of these different mechanisms. Thus the first problem may be said to be one of conductance and the second one of resistance. This leads to an electrical analogy of a set of resistors in parallel and series where the parallel paths represent the different mechanisms of conduction and the size of the resistors the different processes impeding the flow of heat.

An intuitively appealing introduction to theory of thermal conductivity of solids can be obtained by brief consideration of the situation of heat conduction in a gas. Here we have a collection of molecules which possess three significant properties from the point of view of heat conduction. Firstly, as the temperature rises the average energy of the gas molecules increases, so that if molecules from a hot region and from a cold region are exchanged energy will be transferred from the hot to the cold region. Secondly, because the energy of the molecules is largely kinetic, the velocities of the molecules in a hot region are greater than in a cold region. This means that the process of thermal diffusion will be observed by which molecules tend to move from the hotter to the colder region. This is the mechanism of heat transport in a gas. The third essential property is that the molecules collide with each other exchanging energy and momentum, so that energy brought in to a colder region by an energetic molecule is rapidly shared with its fellows, thus maintaining a local thermal equilibrium. An equally important effect of the collisions is to prevent direct energy transfer by a beam of molecules, rather as energy is transferred by radiation in a beam of light. Such an energy transfer will not exhibit the proportionality to temperature gradient, which as indicated in chapter 1 is a necessary property of true heat conduction.

The effect of these collisions is described by a mean free path ℓ, appropriately defined so that after traversing this distance the average molecule comes to equilibrium with those around it. It is shown in the kinetic theory of gases that the thermal conductivity of a gas is given approximately by

$$\lambda = \tfrac{1}{3}C_v\bar{u}\ell \quad (4.1)$$

where C_v is the heat capacity at constant volume per unit volume, and $\bar{u}$ is the average velocity of the molecules. A similar picture provides a model for heat conduction in solids by electrons and more surprisingly by atomic vibrations as well.

The major part of the interpretation of experimental data will be given in chapter 5. However, it is desirable before entering into a discussion of the theory of thermal conductivity to present the essential features of what is observed in experiments on rather simple solids. The simplest case to start with is the insulating single crystal, and a suitable example would be provided by potassium chloride (KCl). Figure 4.1 shows results published by Walker and Pohl (1963) for pure KCl in the temperature range from 1 to 200 K. Notice that both the thermal conductivity and temperature are plotted logarithmically. The principal feature is that the thermal conductivity has a maximum of about 6 W cm^{-1} K^{-1} at 8 K and decreases at both lower and higher temperatures. Above ~50 K, if the temperature range beyond that shown in the figure is included, the slope on the graph is close to -1 showing that $\lambda \propto T^{-1}$. Below the temperature of the thermal conductivity maximum the slope increases tending toward a value of 3 characteristic of many materials at low enough temperatures; here $\lambda \propto T^3$. Furthermore, the thermal conductivity of single crystals is also affected by sample dimensions: the larger the cross section the larger the thermal conductivity and the lower the temperature at which the T^3 dependence appears.

The magnitude of the thermal conductivity maximum and the temperature at which it occurs are strongly influenced by the presence of foreign atoms (impurities or deliberately added) as illustrated by the second curve in figure 4.1 where a concentration of 5×10^{19} cm^{-3} of chlorine has been replaced by iodine giving a very dilute solution of KI in KCl. This has a very considerable effect reducing the maximum thermal conductivity by a factor of four and producing significant, if lesser, reductions at higher and lower temperatures. Clearly the problem of the effects produced by foreign atoms is a very important one.

Because potassium chloride and other materials showing similar behaviour are electrically insulating, it is natural (and correct) to assume that conduction of heat by mobile electrons may be neglected. In the case of metals showing high electrical conductivity such an assumption clearly may not be made and a highly significant result was discovered quite early in the study of heat conduction in metals. This was that for a number of metals at room temperature the ratio of thermal to electrical conductivity is approximately the same. This is called after its discoverers the Wiedemann–Franz law. The further significant fact that the constant itself was proportional to temperature was discovered by Lorenz. This was regarded as establishing that the processes of electrical and thermal conduction were due to the same mechanism involving mobile free electrons.

There are a number of ways in which metals deviate from this simple behaviour. Firstly, at fairly low temperatures the Wiedemann–Franz law often breaks down quite severely followed by a resumption of its sway at even lower temperatures. This is understood in terms of the different effectiveness of various scattering mechanisms for electrical and thermal conductivity. These effects are found in typical pure noble metals of high electrical conductivity. In the case of metals that are less good electrical conductors or of alloys the thermal conductivity exceeds the value suggested by the Wiedemann–Franz law because the lattice thermal conductivity is no longer relatively negligible. Finally, the behaviour of metals which become superconducting suggests that the electronic thermal conductivity starts to disappear when this occurs, leaving only lattice conduction.

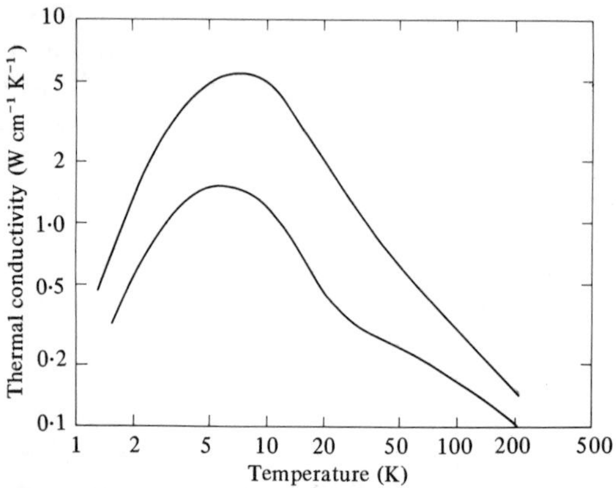

Figure 4.1. General form of the variation of thermal conductivity with temperature of an insulator, illustrated with reference to pure and doped potassium chloride (Walker and Pohl, 1963).

4.2 Lattice vibrations

The idea that the conduction of heat in a solid is due to the interaction of the motion of the atoms that make the solid up dates back to the earliest time at which ideas of the atomic constitution of matter and of heat as motion were current. An essential step towards understanding the transport of heat by this process can be made by considering the nature of the motion of the atoms in a solid, bound together as they are by interatomic forces which, although requiring quantum mechanics for a full explanation, can nevertheless be treated largely in a classical manner.

We suppose initially that the interatomic forces may be described in terms of superimposed attractive and repulsive forces. Without the latter a system of atoms would just collapse and without the former there would be no tendency to condense at all. In the simplest fashion these forces may be supposed to result in the location of the atoms in a regular array or lattice, the different arrangements comprising the field of study of crystallography. Such an arrangement of atoms will clearly be able to vibrate, and because of the enormous numbers of atoms these vibrations might be supposed to be very complicated. They can be simplified, however, in the same way as any other small vibrations, by analysis into normal modes. Each normal mode has its own characteristic frequency and characteristic atomic movements which can occur completely independently of any other normal mode. Any realisable vibration, however complex, can be regarded as a superposition of normal-mode vibrations, this process being analogous to Fourier analysis. In the case of a crystalline solid the crystal symmetry itself gives rise to a great simplification, because the normal modes can be described as either standing or travelling waves (the two descriptions are equivalent). Thus in such solids the normal modes are characterised not only by a particular frequency ω but also by (i) the wave vector $\boldsymbol{q}$, whose magnitude q is the reciprocal of the wavelength times 2π and whose direction is the direction of propagation, and (ii) the polarisation vectors giving the direction of the atomic displacements. We shall use the label s to specify polarisation. It is customary to regard frequency as a function of $\boldsymbol{q}$ and s: $\omega = \omega_s(\boldsymbol{q})$.

At long wavelengths where the frequency is low the lattice vibrations can be identified with the sound waves in the crystal so that if v_s is the velocity of sound for a wave of polarisation s we shall have

$$\omega = v_s q. \tag{4.2}$$

As the wavelength decreases this simple relationship breaks down ω usually falling below $v_s q$. The resulting dependence of ω on q, called a dispersion relation, can be determined experimentally and is illustrated in figure 4.2. In simple cases the polarisation may be described as longitudinal or transverse and the former always has the higher frequency. The figure also shows that only a certain range of q is allowed. In the one-dimensional case shown there, $-\pi/a < q < \pi/a$ where a is the unit cell dimension. The reason for this is that motion corresponding to larger values of q turns out to be the same as that given by some value of q lying within the allowed range. Similarly in three dimensions all the normal modes can be represented by points in '$\boldsymbol{q}$-space' contained within a certain volume called the Brillouin zone. It is in fact found that the number of allowed values of q and s is just such as to provide three normal modes per atom in the solid, this being the total number of degrees of freedom of the system, $3N$ for N atoms.

If there is more than one atom in each periodic unit (cell) in the crystal, these dispersion curves become more complicated. Figure 4.3 shows the situation with two atoms in the cell. The modes of vibration corresponding to the upper curves are called optical because their frequencies are in the infrared range; the others whose frequencies vanish as $q \rightarrow 0$ are called acoustic modes because of their connection with the propagation of sound. Each of the different branches of the dispersion relation is labelled with a different s, so in the case shown in figure 4.3 there are six values of s labelled in order of increasing frequency TA1, TA2, LA, TO1, TO2, and LO. Here L stands for longitudinal, T for transverse, A for acoustical and O for optical, whilst the two transverse branches whose frequencies are very similar are distinguished by the numbers 1 and 2.

Let us suppose that the set of normal modes required for a complete specification of the possible vibrations of a solid has been found; the next important question is how energy would be distributed over these modes in thermal equilibrium. The first stage in finding the answer is to note that each normal mode is, in fact, a harmonic oscillator, and that the quantum mechanics of such oscillators specifies that only certain values of energy are allowed:

$$\mathcal{E} = (n+\tfrac{1}{2})\hbar\omega \tag{4.3}$$

where n is an integer (0, 1, 2, ...) and $\hbar$ is Planck's constant divided by 2π. The $\frac{1}{2}\hbar\omega$ term is the zero-point energy and does not usually play any significant part. The remaining energy is said to be in the form of n 'phonons', a name derived by analogy with the photons of radiation in the

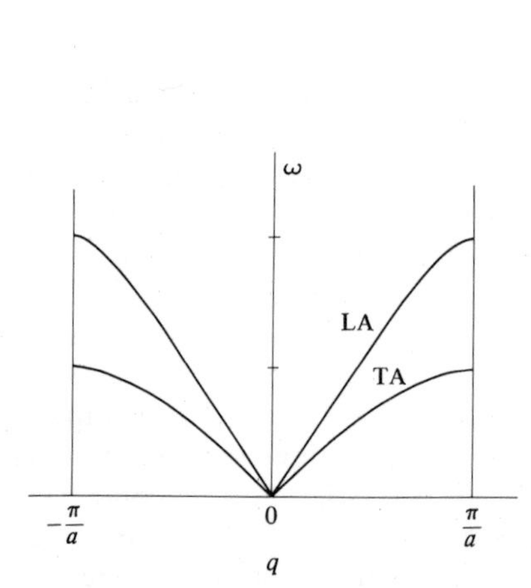

Figure 4.2. Dispersion curves for lattice vibrations where there is one atom in the unit cell.

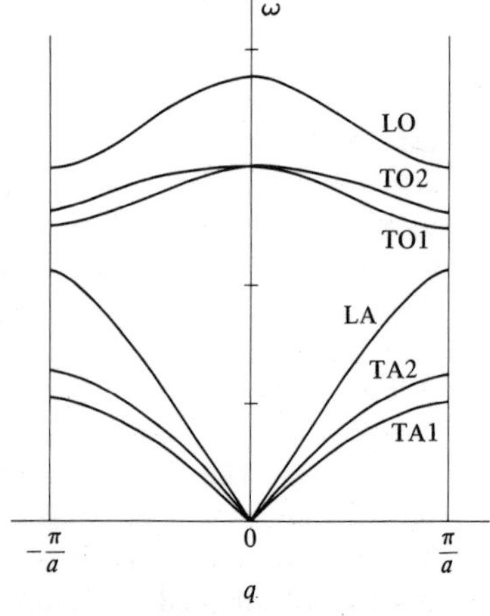

Figure 4.3. Dispersion curves for lattice vibrations where there are two atoms in the unit cell.

electromagnetic field. It will be seen that in the event of a change in the energy of a normal mode, this can only occur in units of $\hbar\omega$, i.e. in phonons. To give a precise definition, the phonons are the quanta of excitation of the lattice vibrational normal modes.

* The phonon concept turns out to be an extremely valuable one, because it provides an intuitive picture of lattice vibrational heat transfer, based on an analogy with the molecules of a gas. One has two equivalent pictures: first, of a set of normal modes, each excited to a particular allowed level of energy specified by the quantum number; and second, of a set of phonons, having many of the properties of particles, the number of phonons of a given energy being given by the quantum number used in the mode picture. Thus the vibrational motion of an enormous number of atoms is first described in terms of a set of normal modes, which is then replaced by a gas of phonons. It is easy to see how it became common usage to speak of heat conduction by phonons rather than by lattice vibrations, the notion having been borrowed from the kinetic theory of heat conduction in gases.

It is necessary to stress the peculiar properties of the phonons regarded as particles. Like a particle they have momentum $\hbar\boldsymbol{q}$ as well as energy $\hbar\omega$. On the other hand they have no mass nor are they conserved in the way in which atoms or electrons are. The total number of phonons is not fixed, although their total energy is. Finally, all phonons of the same mode are indistinguishable. This is usual for elementary particles and has an important bearing on the statistics they obey.

The next stage is in fact to use the appropriate statistics to calculate the average value of the quantum number in thermal equilibrium, i.e. to calculate the average number of phonons of a given type. The result is given by the Planck function

$$\tilde{n} = \left[\exp\left(\frac{\hbar\omega}{k_B T}\right) - 1\right]^{-1}, \qquad (4.4)$$

where k_B is Boltzmann's constant and T is the thermodynamic temperature. The same function describes the energy distribution in black body radiation. If the quantity $\hbar\omega/k_B T$ is very small then

$$\tilde{n} \approx \frac{k_B T}{\hbar\omega}.$$

Since the energy of the phonon is $\hbar\omega$, this means that the average energy per normal mode is $k_B T$. This result expresses the classical equipartition of energy and leads directly to the Dulong and Petit law which states that at high temperatures the molar heat capacity of a solid is equal to $3N_A k_B \approx 25\ \mathrm{J\ K^{-1}}$, N_A being Avogadro's number.

It is in the calculation of specific heat that the theory of lattice vibrations finds its most direct application. To see how this is done, one first writes down the vibrational energy of the solid

$$E = \sum_{s,q} \tilde{n}\hbar\omega_s(\boldsymbol{q}) \tag{4.5}$$

as the sum of the average energy of the individual modes (q, s). The next step is to replace the sum by an integral over ω. To do this we require the density of states function $g(\omega)$, which is defined as the number of normal modes with frequencies between ω and $\omega+d\omega$. One then has

$$E = \hbar\int_0^\infty g(\omega)\tilde{n}\omega d\omega. \tag{4.6}$$

It is necessary that the total number of modes is $3N$, so

$$\int_0^\infty g(\omega)d\omega = 3N. \tag{4.7}$$

Since we know the form of $\tilde{n}$, the main problem in determining E is to find $g(\omega)$. One particularly simple form corresponds to the assumption that all the modes have the same frequency: this was the basis of Einstein's original work on specific heats. The most useful form of $g(\omega)$ is undoubtedly that suggested by Debye in his theory of specific heats. Debye assumed that all the lattice vibrational modes were in fact of the nature of sound waves so that ω was given by equation (4.2). With this assumption the constant frequency surfaces in $\boldsymbol{q}$-space become spheres and the number of modes with less than a certain frequency is proportional to the volume of the sphere. In fact,

$$\int_0^\omega g(\omega')d\omega' = 3\frac{V}{8\pi^3}\frac{4\pi}{3}\left(\frac{\omega}{v}\right)^3,$$

where all polarisations have been assumed to have the same velocity v and $V/8\pi^3$ is the density of allowed values of $\boldsymbol{q}$ in $\boldsymbol{q}$-space for a crystal of volume V. Clearly,

$$g(\omega) = \frac{3V}{2\pi^2}\frac{\omega^2}{v^3}. \tag{4.8}$$

In order to give the correct total number of modes in equation (4.7) Debye simply cuts off $g(\omega)$ above an upper limit ω_D, so

$$g(\omega) = 0, \qquad \omega > \omega_D, \tag{4.8a}$$

and using (4.7),

$$\frac{V}{2\pi^2}\left(\frac{\omega_D}{v}\right)^3 = 3N,$$

determines ω_D. Now the internal energy becomes from (4.6)

$$E = \frac{3V\hbar}{2\pi^2 v^3}\int_0^{\omega_D} \omega^3\left[\exp\left(\frac{\hbar\omega}{k_B T}\right)-1\right]^{-1} d\omega, \tag{4.9}$$

and the specific heat is just $\partial E/\partial T$.

If we set $\hbar\omega/k_B T = z$ then this equation takes the form

$$E = 9Nk_B T\left(\frac{T}{\Theta_D}\right)^3 \int_0^{\Theta_D/T} \frac{z^3}{e^z - 1}dz, \tag{4.9a}$$

where $\Theta_D = \hbar\omega_D/k_B$ defines the Debye temperature.

At low temperatures $\tilde{n}$ becomes very small at frequencies above ω_D so it is not a bad approximation to replace the upper limit in (4.9) with infinity. Inspection of E then shows it to be proportional to T^4, so the specific heat varies as T^3, this being the famous Debye T^3 law.

Debye's theory has proved to be immensely successful despite the fact that the form of $g(\omega)$ given by expression (4.8) is often very wide of the mark. This causes some relatively minor discrepancies at low temperatures. Furthermore, this theory is the basis of almost all discussion of phonon thermal conductivity.

It is not difficult to write down the heat carried by a gas of phonons. Each phonon has an energy $\hbar\omega$ and a velocity v_s in the direction of $\boldsymbol{q}$ so it contributes $\hbar\omega_s(\boldsymbol{q})v_s\boldsymbol{q}/q$ to the heat current. Thus the total heat current is

$$U = \hbar \sum_{s,\boldsymbol{q}} \omega_s(\boldsymbol{q})v_s\frac{\boldsymbol{q}}{q}N_s^q, \tag{4.10}$$

where N_s^q is the average number of phonons belonging to mode $(\boldsymbol{q}, s)$. If N_s^q is given by the Planck function $\tilde{N}_s^q$, $\boldsymbol{U}$ will vanish because $\omega_s(-\boldsymbol{q}) = \omega_s(\boldsymbol{q})$ and $\tilde{N}_s^q$ depends only on $\omega_s(\boldsymbol{q})$. It is of course necessary on thermodynamic grounds that the heat current vanishes in equilibrium. Obviously (4.10) is not of much use as it stands, because N_s^q is unknown, and the greater part of the theory of lattice thermal conductivity is concerned with the determination of N_s^q. This we shall return to later.

The description of lattice vibrations given above is intended only as an introduction to the subject. The results quoted here are derived in many textbooks, among which that of Kittel (1970) is in many ways outstanding. More advanced accounts of lattice vibrational theory are to be found in the works of Ziman (1960) and Maradudin *et al.* (1971).

4.3 Electrons in solids

It was the discovery of Wiedemann and Franz which suggested that the same mechanism which provided electrical conduction also gave conduction of heat. At the beginning of this century theoretical models, based on a gas of mobile 'free' electrons moving through a lattice of positive ions proved capable of explaining the law of Wiedemann and Franz in a semi-quantitative fashion. However, these models left much to be desired. For instance classical ideas suggested that free electrons should contribute to the specific heat of a metal as much as monatomic molecules contribute to that of a gas, but in fact the lattice vibrations seemed capable of accounting for the whole of the observed specific heat. Again, the mean free path of the electrons might have been expected to approximate to the interionic spacing in the metal, but experimentally appeared to be much larger. These anomalies all require quantum mechanics for their explanation.

One can say straightaway that the present theoretical picture seems perfectly capable of explaining the broad features of the experimental situation, although the detail is extraordinarily complicated in many respects. A distinctly simplified account will be given which, however, can account for all but the finest structure in the electronic part of heat conduction.

Quantum mechanics shows that in solids there are certain states available for electrons to occupy and that these states are specified by a wave vector $\boldsymbol{k}$ and an energy $\mathcal{E}$. The allowed values of $\boldsymbol{k}$ are the same as those of $\boldsymbol{q}$ in the lattice vibration case, and for a given value of $\boldsymbol{k}$ there will be many values of energy so that $\mathcal{E}$ can be regarded as a function of $\boldsymbol{k}$ and a 'band index' r: $\mathcal{E}_r(\boldsymbol{k})$. The term 'band index' is used because the energy levels are grouped in quasi-continuous bands with gaps between them where there are no permitted energy states. This is illustrated in the one-dimensional case in figure 4.4. Here the energy gap must be absolute, but in the more realistic case of three dimensions it may be that, although in one direction there may be no allowed levels, in a particular energy range there will be levels in this range in a different direction in '$\boldsymbol{k}$-space'.

Thus a major problem in the theory of solids is the determination of $\mathcal{E}_r(\boldsymbol{k})$, just as in the lattice vibration case much importance attaches to finding $\omega_s(\boldsymbol{q})$ for the normal modes. If $\mathcal{E}_r(\boldsymbol{k})$ is known then a density of states function $N(\mathcal{E})$ can be determined such that the number of energy levels between $\mathcal{E}$ and $\mathcal{E}+\mathrm{d}\mathcal{E}$ is $N(\mathcal{E})\mathrm{d}\mathcal{E}$. This is analogous to $g(\omega)$ in the

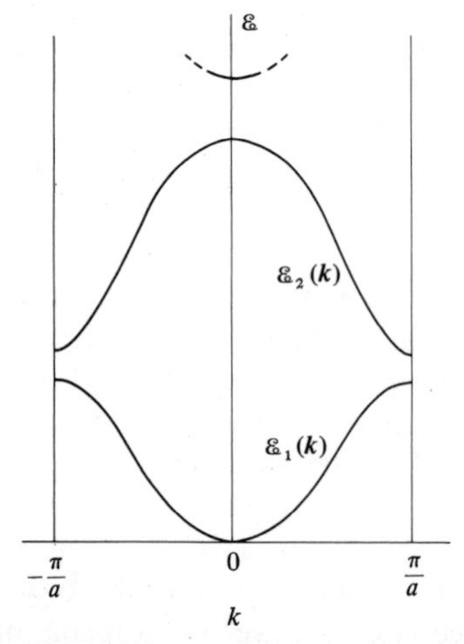

Figure 4.4. Simplified form of energy bands in a solid.

case of lattice vibrations. Different possible forms of $N(\mathscr{E})$ are shown in figures 4.5, 4.6 and 4.7. Many properties of solids can be described largely by $N(\mathscr{E})$ provided the statistical distribution of the electrons over their allowed levels is also known.

At absolute zero this distribution is simply determined by the Pauli principle which states that only one electron is allowed in each level (actually each level is double owing to the two allowed spin orientations for the electrons). The number of electrons is determined by the condition of electrical neutrality or, put another way, by the number of electrons in the free atoms from which the solid may be supposed to have condensed. Thus at absolute zero there will be an energy below which all the states are filled and above which they are all empty. This energy is called the Fermi energy and is denoted by $\mathscr{E}_F$. This result forms the basis of a very important distinction between two types of solids. If the top of the filled energy levels corresponds to the bottom of an energy gap, then we have an insulator if the gap is large (say greater than 2 eV) or a semiconductor if it is small (figure 4.5). If, on the other hand, the highest filled level has a vacant allowed level immediately above it, we have either a metal if $N(\mathscr{E})$ is large (figure 4.6) or a semimetal if $N(\mathscr{E})$ is small (figure 4.7).

The reason for this is that a filled band cannot conduct electricity, so that in the case of an insulator or semiconductor a considerable amount of energy is necessary to excite an electron from the filled band to the

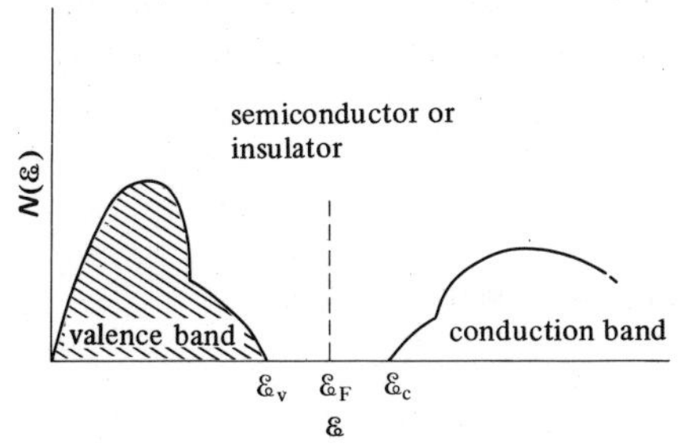

Figure 4.5. Electronic density of states function for a semiconductor or an insulator.

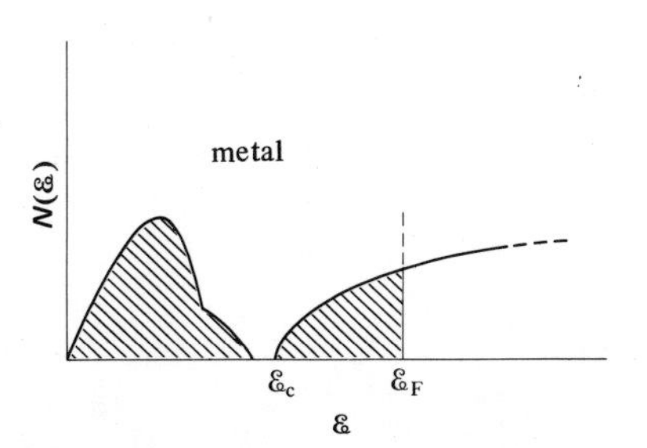

Figure 4.6. Electronic density of states function for a metal.

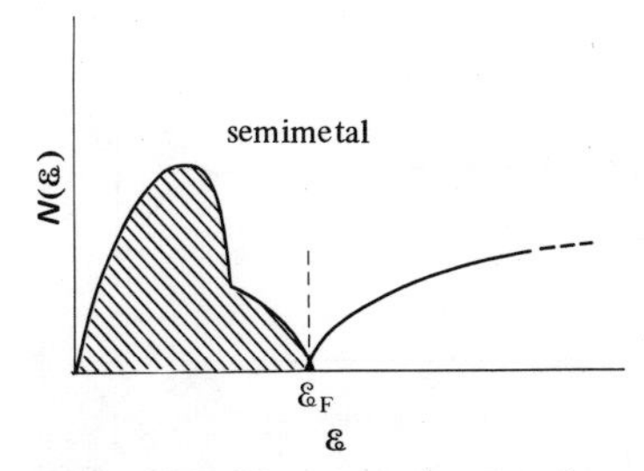

Figure 4.7. Electronic density of states function for a semimetal.

nearest empty band in order that conduction can occur. However, if this does occur, not only does the electron in the higher band conduct, but also the vacant state in the lower band acts as a mobile positive charge called a 'hole'.

Energy will be available at all temperatures above absolute zero. The probability of finding an electron in a level at energy $\mathcal{E}$ is given by the Fermi function

$$f_0 = \left[\exp\left(\frac{\mathcal{E} - \mathcal{E}_F}{k_B T}\right) + 1\right]^{-1} \tag{4.11}$$

plotted in figure 4.8. If $\mathcal{E} - \mathcal{E}_F \gg k_B T$ then

$$f_0 \approx \exp\left(\frac{\mathcal{E}_F - \mathcal{E}}{k_B T}\right), \tag{4.12}$$

which is of the Maxwell–Boltzmann form. The probability of finding a vacant level, i.e. a hole, is given by $1 - f_0$; if $\mathcal{E}_F - \mathcal{E} \gg k_B T$ this is given by

$$1 - f_0 \approx \exp\left(\frac{\mathcal{E} - \mathcal{E}_F}{k_B T}\right). \tag{4.13}$$

Thus, if $\mathcal{E}_F$ lies inside an energy gap, equations (4.12) and (4.13) show that the electrons and holes will behave like classical particles, at least as far as statistics are concerned.

We are now in a position to write down an expression for the total energy of the electrons in a fashion analogous to equation (4.6) for phonons:

$$E = \int_0^\infty N(\mathcal{E}) f_0 \mathcal{E} \, d\mathcal{E}. \tag{4.14}$$

Analogous to (4.7) is the equation determining $\mathcal{E}_F$:

$$\int_0^\infty N(\mathcal{E}) f_0 \, d\mathcal{E} = N, \tag{4.15}$$

where N is the total number of electrons.

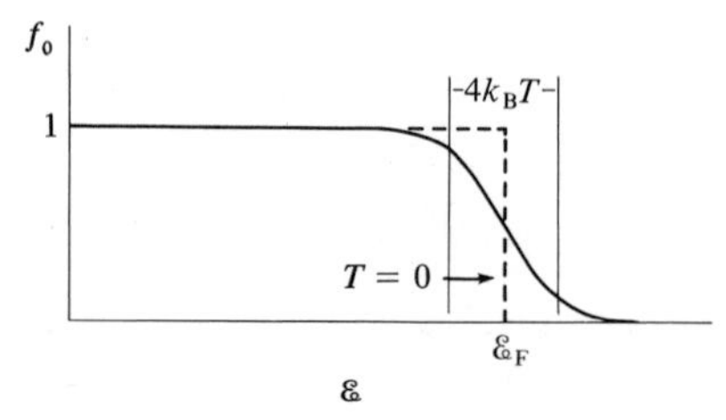

Figure 4.8. The form of the Fermi function f_0.

The electronic specific heat is given by $\partial E/\partial T$. If the value of $\mathscr{E}_F$ is such that $N(\mathscr{E}_F)$ is large then, because of the form of $\partial f_0/\partial T$ shown in figure 4.9, only a small proportion of the total number of electrons contribute to the specific heat. If, on the other hand, $\mathscr{E}_F$ lies in an energy gap then, again, there are only very few free electrons and holes compared with the number of atoms in the solid, so again the electronic specific heat is negligible. This removes one of the major stumbling blocks in the early free-electron theory of solids.

A model based on the similarity of lattice vibrations to sound waves has proved extremely useful. Similarly, in the case of electrons, the effective mass approximation is of great value. This treats the electrons as if they were free electrons but with a mass m_e^* differing from the free electron value. This means that

$$\mathscr{E} = \frac{\hbar^2 k^2}{2m_e^*} \tag{4.16}$$

and

$$N(\mathscr{E}) = \frac{1}{2\pi^2}\left(\frac{2m_e^*}{\hbar^2}\right)^{3/2} \mathscr{E}^{1/2}. \tag{4.17}$$

A similar model can be used for holes with a hole effective mass m_h^*. In semiconductor theory m_e^* applies to the lowest nearly empty band and m_h^* to the highest nearly filled band. In reality m_e^* (and m_h^*) may be anisotropic and also a function of energy, and indeed in some cases becomes so complicated that it loses its usefulness entirely.

An important concept in the case of metals is that of the Fermi surface. If the simple model of (4.16) is used, the Fermi energy $\mathscr{E}_F$ defines a sphere in $\boldsymbol{k}$-space within which all the electronic levels are full, at least at absolute zero. Thus we have a spherical Fermi surface. In real metals we also have a Fermi surface which is no longer spherical but, within limits imposed by symmetry, sometimes takes extremely fantastic forms; though it is surprising how often the Fermi surface can be shown to be a sphere on which some distortion has been imposed.

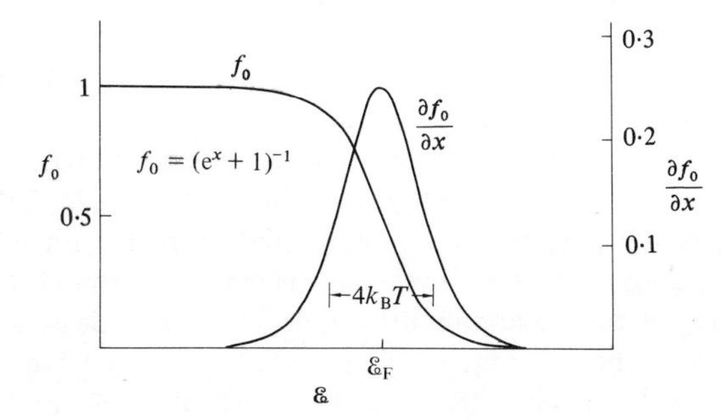

Figure 4.9. The form of the Fermi function and its derivative.

One important consequence of the wavelike properties of electrons was that the periodic array of positive ions in a perfect crystal would not scatter the electrons at all and thus not contribute to electrical and thermal resistance. Only departures from regularity, such as lattice vibrations or foreign atoms, would do this. Scattering mechanisms of this kind will be of considerable concern to us in later sections.

The most drastic change that can occur in the electronic behaviour of solids is the onset of superconductivity. From a phenomenological standpoint the situation can be described by saying that a certain proportion of the electrons go into a state of zero entropy separated by an energy gap from the normal electronic levels. This proportion increases as the temperature decreases. In this zero entropy state the electrons are able to carry an electric current without resistance, but there can be no heat transport or interaction with phonons or crystal defects. This has very important consequences for the thermal conductivity.

Just as in the case of lattice vibrations this account of electrons in solids is only an introduction. For further reading Kittel's (1970) book is again excellent; a more advanced account from a modern standpoint is given by Harrison (1970).

4.4 Lattice heat conduction

It has been suggested earlier [equation (4.1)] that the lattice thermal conductivity can be adequately discussed in terms of a phonon gas having a particular specific heat and mean free path. In this section the question will be considered from a more fundamental point of view. The central concept here is that of a Boltzmann equation (Ziman, 1960) which shows how it may be possible to go from the formal expression for a heat current, equation (4.10), to a usable expression for the thermal conductivity by determining N_s^q, the phonon distribution function. If we suppose that we can generalise the quantity N_s^q so that it can be a function of position $\boldsymbol{r}$ as well as of wave vector $\boldsymbol{q}$ and polarisation s then the Boltzmann equation is a gain–loss equation describing the way in which $N_s^q(\boldsymbol{r})$ can change owing to the presence of a temperature gradient and the existence of scattering processes for phonons. If the reader ponders the way in which N_s^q was originally defined with reference to normal modes belonging to the whole of crystal, he will realise that the derivation of the Boltzmann equation itself is by no means straightforward.

This problem is not particularly difficult in classical physics where the Boltzmann equation was originally employed in the kinetic theory of gases. For many years the phonon Boltzmann equation rested on not much more than an analogy, since in quantum mechanics the uncertainty principle prevents the use of the arguments employed in the classical case. However, in recent years the techniques of many-body theory have enabled more rigorous proofs to be constructed as in the work of Kwok and Martin (1966), Kwok (1967), and Horie and Krumhansl (1964). These authors have not only derived the usual Boltzmann equation but also shown what corrections might be applied if necessary. In addition, the many-body theory provides through the Kubo method (see e.g. Schieve and Peterson, 1962, and Maradudin, 1964) a direct approach to the thermal conductivity without using a Boltzmann equation, although this has not proved very convenient in practice, beyond proving perhaps more rigorously the validity of the kinetic theory approach. It is reassuring to know that what had been done for many years on an almost entirely intuitive basis was in fact on sound foundations.

The validity of the Boltzmann equation having been established, the question is then how to solve it. There are basically two approaches to this. If we can write the rate of change of the distribution function N_s^q due to collisions and other scattering processes as

$$\left(\frac{\partial N_s^q}{\partial t}\right)_{\text{coll}} = \frac{\widetilde{N}_s^q - N_s^q}{\tau_s^q} ,$$

then we have a well defined relaxation time τ_s^q (related to the mean free path by $\ell = \tau v$) and the Boltzmann equation is easily solved leading to equation (4.1) in the form of an integral. Unfortunately in the case of phonons it is frequently not possible to derive a well defined relaxation time, which requires us to consider the second approach. This uses the fact that the Boltzmann equation can be shown to be equivalent to a variational theorem, whose physical content may be expressed by saying that matters arrange themselves so that the rate of entropy generation is a minimum. The principles of this method have been discussed thoroughly by Drabble and Goldsmid (1962) and Ziman (1960). There are a number of practical disadvantages which have discouraged the use of the variational method; in particular it is difficult to derive sufficiently simple mathematical expressions for use in the analysis of experimental data. Recent developments which should make its use more practicable have been described by Benin (1970) and Srivastava (1973). However, most theoretical effort has gone into refining the former approach so as to make it possible to calculate effective relaxation times and overcome the N-process problem which was for many years the principal bugbear in this field. This problem will be discussed in detail later on. Klemens (1958) and Carruthers (1961) have reviewed the relaxation time method.

If the variational method is used then the different phonon scattering mechanisms are normally incorporated in the form of transition probabilities which are calculated with the use of perturbation theory. For scattering by a static defect these will be written in the form $Q_{qs}^{q's'}$, so that in a time dt the probability that a (q,s) phonon will be scattered to become a (q',s') phonon is $Q_{qs}^{q's'}\,dt$. These transition probabilities are also used to calculate the relaxation times in that form of the theory. For an elastic scattering process the relaxation time τ_s^q will be given by an equation of the form

$$\frac{1}{\tau_s^q} \approx \sum_{q's'} (1 - \cos\vartheta) Q_{qs}^{q's'} \tag{4.18}$$

where ϑ is the angle between $\boldsymbol{q}$ and $\boldsymbol{q}'$, so the term $(1 - \cos\vartheta)$ allows for the fact that scattering through a small angle is less effective than through a large angle. A large part of the theory consists of calculations of transition probabilities and, where possible, relaxation times for all significant scattering processes. The most important are phonon–phonon scattering involving three phonons, scattering by point defects, and at external or internal boundaries. In the case of metals, scattering of phonons by electrons is very important.

The conditions for the existence of a well-defined relaxation time are that the scattering must be elastic and that the scattered phonon must be of the same polarisation type (s) as the initial phonon. The first condition manifestly does not hold for processes involving three phonons in which, for example, two phonons combine to give a third. Such processes are predominant in all fairly pure nonmetals at room temperature and above, so this is a very considerable limitation. However, the first condition is satisfied for all scattering by static defects such as impurity atoms. Unfortunately, the second condition is not satisfied in the large proportion of these latter events in which, for example, a transverse phonon is scattered into a longitudinal phonon. One way of evading this difficulty is to assume that in place of a true relaxation time it is possible to substitute a so-called single-mode relaxation time. The reason why a proper relaxation time has to satisfy the conditions specified earlier is that all the phonons (or normal modes) are disturbed from equilibrium by the existence of a temperature gradient. If we assume that only one mode is disturbed and all the others are in equilibrium, then there is no difficulty in writing down an expression which would describe in the form e^{-t/τ^*} the return of that mode to equilibrium. τ^* is then the single-mode relaxation time and will take the same form as τ in equation (4.18) without the $\cos\vartheta$ term. In many cases it is likely that $\tau^* \approx \tau$, and a discussion of this point is given by Carruthers (1961) and others. Unfortunately this is least plausible for the three-phonon processes at room temperature and above, but in order to make headway τ^* is commonly used even in this situation.

The lattice thermal conductivity λ_{ph} can be written down by finding N_s^q from the Boltzmann equation, putting it in equation (4.10) and dividing by the negative of the temperature gradient. The resulting expression containing the relaxation time is still quite complex but, if it is assumed that ω is a function of the magnitude of $\boldsymbol{q}$ but not its direction, a relatively simple form is obtained:

$$\lambda_{ph} = \frac{\hbar^2}{3k_B T^2} \sum_s \int \omega^2 v_s^2 \tau_s \tilde{n} (\tilde{n} + 1) g_s(\omega) d\omega, \tag{4.19}$$

where all the quantities in the integrand may be functions of ω. $g_s(\omega)$ is the density of states function for a particular polarisation s. The above equation is based on the assumption that the substance is isotropic. This is not generally true, but the consequences of anisotropy have rarely been considered in theoretical discussion, almost all of which is based on the Debye approximation, where

$$g_s(\omega) = \frac{\omega^2}{2\pi^2 v_s^3}\,, \tag{4.20}$$

provided $\omega < \omega_D$ and equals zero for frequencies exceeding ω_D. [The absence of V in equation (4.20) compared with (4.8) is due to the fact that we now wish $g_s(\omega)$ to refer to unit volume.] The analogy with equation (4.1) may be brought out by noticing that, since the heat capacity of a mode is

$$C = \frac{\hbar^2\omega^2\widetilde{n}(\widetilde{n}+1)}{k_B T^2}$$

and its mean free path is $\ell_s = v_s\tau_s$, the right hand side of equation (4.19) is an integral over the quantity $\frac{1}{3}Cv_s\ell_s$ weighted by the density of states function $g_s(\omega)$ and then summed over all polarisations s. If there are a number of processes, each described by a relaxation time τ_s^i, their combined effect can be obtained by reciprocal addition:

$$\frac{1}{\tau_s} = \sum_i \frac{1}{\tau_s^i}\,. \tag{4.21}$$

It will be seen that in order for it to be possible to use equation (4.19) it is desirable that the relaxation times should be known as functions of the angular frequency ω. Finally, it may be necessary to correct (4.19) to allow for the difficulty over N-processes. The next step will now be to consider the individual scattering mechanisms.

4.5 Phonon scattering processes

The most important and, unfortunately, the most complex scattering mechanisms are those involving three phonons; they are of two kinds. In the first, A, two phonons combine to form a single phonon, while in the second, B, one phonon breaks up to give two phonons. These processes are shown schematically in figure (4.10), where the phonon mode labels are also shown.

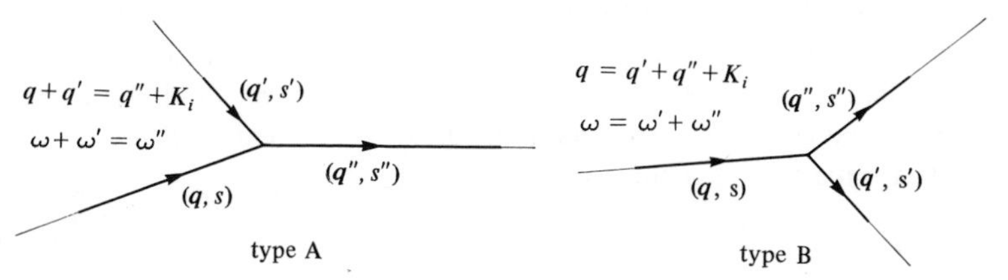

Figure 4.10. Type A and type B three-phonon processes.

The physical basis of these processes lies in the anharmonic (third order) terms in the lattice potential energy. This can be seen most clearly by reverting to the wave picture at long wavelengths. Here the anharmonic effects show themselves as a strain dependence of the elastic constants and therefore of the velocity of sound. Thus a sound wave will produce a periodic variation in the sound velocity in a medium and this will perturb the passage of a second sound wave. Similarly, the second wave perturbs the first giving a symmetrical interaction. Now this is analogous to two periodic oscillations imposed simultaneously on a nonlinear system which, it is well known, produces sum and difference frequency oscillations at the output of the system. The sum frequency corresponds to process A and the difference frequency to process B. Because the nonlinear interaction occurs over the volume of the medium there will also be relations involving the wave vectors. The best analogy here is that the first sound wave acts as a diffraction grating for the second so that there will be interference conditions involving $\boldsymbol{q}$, $\boldsymbol{q}'$, and $\boldsymbol{q}''$.

The origins of the interaction having been established, it is now more convenient to return to the phonon picture. Assuming that such an interaction is possible, one will clearly expect that there will be energy and momentum conservation in the three-phonon processes. Energy conservation will mean that for process A

$$\omega+\omega' = \omega'', \tag{4.22a}$$

and for process B

$$\omega = \omega'+\omega'', \tag{4.22b}$$

where the energy has been divided by $\hbar$ and the mode labels $(\boldsymbol{q},s)$ etc. have been dropped. These equations correspond to the sum and difference frequencies in the wave picture. Momentum conservation will be established by means of an equation like

$$\boldsymbol{q}+\boldsymbol{q}' = \boldsymbol{q}''$$

for the first type of process, where this time the momenta have been divided by $\hbar$. This equation also expresses the interference condition for the wave picture. However, this is not the whole story, because till now the argument has neglected the atomic nature of the medium and in particular the existence of a lattice. Keeping with the wave picture, the lattice itself acts as a diffraction grating in the manner familiar in x-ray crystallography. The character of the lattice is described by a set of 'reciprocal lattice translation vectors' denoted by $\boldsymbol{K}_i$, and the correct interference condition must take account of it. This is done by modifying the last equation into the form

$$\boldsymbol{q}+\boldsymbol{q}' = \boldsymbol{q}''+\boldsymbol{K}_i, \tag{4.23a}$$

where K_i is one of the reciprocal lattice translation vectors, including one with zero value. Process B gives a condition of the form

$$q = q' + q'' + K_i \,. \tag{4.23b}$$

If $K_i = 0$, the three-phonon scattering is called a normal process or N-process; if $K_i \neq 0$, it is called an umklapp process or U-process. These names were given to these interactions by Peierls in his pioneering paper on thermal conductivity. It is perhaps worth commenting that a detailed study of the consequences of equations (4.22) and (4.23) shows that only the smallest reciprocal lattice translation vectors can take part in the three-phonon scattering and that the magnitude of these vectors is approximately $2\omega_D/v$.

Let us consider a gas of phonons where the only scattering is by N-processes. Since these processes are momentum conserving, the total momentum of the phonon gas will be unchanged by such interactions, and once a current of phonons had been established it would persist indefinitely. Such a current corresponds, of course, to a flow of heat, so that we have found that in this situation such a flow of heat can continue without a temperature gradient being required to sustain it. Thus N-processes alone cannot produce a nonvanishing thermal resistance. A parallel may be drawn with the situation in a gas of molecules capable only of momentum-conserving collisions with one another. The flow of such a gas will clearly not be impeded by collisions like these. This is the crux of what was referred to earlier as the N-process problem.

It might be supposed that the problem could be easily solved by ignoring the N-processes completely in calculating the thermal resistivity, but this is not the case. The reason is that the N-processes frequently enhance the effect of the remaining scattering mechanisms. If the phonon gas started with some arbitrary distribution of energy over the different normal modes, then it can be shown that the N-processes redistribute this energy until a particular quasi-equilibrium distribution is achieved, which must, however, have the same total momentum as the initial distribution. Now the non-momentum conserving collisions will generally operate more strongly on some phonons than on others and it frequently occurs that the N-processes shift energy and momentum from phonons which are weakly scattered to phonons which are strongly scattered. In this way the scattering processes in which momentum is not conserved are strengthened by the N-processes.

The N-processes present a real problem only if one attempts to calculate the thermal conductivity using relaxation times, since the variational method takes these processes into account in a fully rigorous manner. However, such are the advantages of the relaxation time method that a number of schemes have been proposed to incorporate the N-processes and we shall return to these later.

A reasonably detailed calculation of the three-phonon processes leads to very complicated expressions and no attempt will be made to reproduce such a calculation here. The complexity is partly physical and partly geometrical. The physical complexity arises in any attempt to portray realistically the anharmonicity of the lattice potential energy. One usable picture relates the anharmonicity to the third-order elastic constants, but commonly it is described by a single parameter called the Grüneisen constant γ, which is rather tenuously connected to a constant of the same name in thermodynamics defined by $-\partial(\log \omega_D)/\partial(\log V)$. The geometrical complexity arises from the necessity of simultaneously satisfying equations (4.22) and (4.23). Fortunately this also results in the elimination of some combinations of polarisations (s, s', s''). U-processes, in particular, pose a considerable problem owing to the necessity of considering various different $\boldsymbol{K}_i$. However, it is possible to arrive at approximate single mode relaxation times, where the dependence of τ on ω is probably fairly reliable in the low frequency limit. These results are summarised in table 4.1.

Table 4.1 categorises the collision processes into three classes. Class I is distinguished by the fact that the three wave vectors, $\boldsymbol{q}$, $\boldsymbol{q}'$, and $\boldsymbol{q}''$, are parallel. In principle there are also collisions involving three transverse phonons, but it can be shown that the strength of these processes is vanishingly small. For many years collisions involving parallel phonons were ignored because of the fact that the slightest departure from the linear proportionality of ω and q prevents them from occurring. However, it was shown by Simons (1963) and Bross (1962) that collision broadening produced sufficient variation of ω to lift this restriction, and for that reason they are included in table 4.1 although there remains some slight doubt about their status.

Table 4.1.

Class	Scattering process	ω and T dependence of τ^{-1}	
		N-processes	U-processes
I	L+L → L	ωT (high temp.) ωT^4 (low temp.)	no U-processes
	L → L+L	$\omega^4 T$	
II	T+L → L	ωT (high temp.) ωT^4 (low temp.)	$\omega^2 T$ (high temp.) $\omega^2 T^{-1}$ $(\exp(-\hbar\omega_D/k_B T)$ (low temp.)
III	L+T ⇌ L T+T ⇌ L	$\omega^4 T$	no U-processes below a minimum frequency

Classes II and III involve nonparallel phonons. In II a low frequency phonon $\boldsymbol{q}$ can interact with phonons $\boldsymbol{q}$ and $\boldsymbol{q}'$ of much larger frequency, but in III only with those of comparable frequency. It is for this reason that low frequency transverse phonons can take part in U-processes, whilst low frequency longitudinal phonons cannot. This also accounts for the strong frequency dependence of the class III relaxation times. The absence of U-processes for low frequency longitudinal phonons suggests that the redistributive role of the N-processes may be highly significant for these modes.

The results presented in table 4.1 should not be regarded as giving the whole story. They were derived on the basis of an isotropic model, and, as Herring (1954) has shown, anisotropy may be very important in producing additional collision processes whose frequency dependence is intermediate between ω and ω^4. Nevertheless, in view of the difficulties attendant on quantitative theories of anisotropic materials, table 4.1 probably provides as much as is practicable for the analysis of experimental data.

As well as three-phonon processes involving only acoustical phonons, it is possible to have processes where one phonon is optical, which will take the form $\mathrm{A}+\mathrm{A} \rightleftharpoons 0$. Such an interaction can only occur if the frequency of the optical phonon is not greater than twice the highest acoustical frequency, and for this it is necessary that the ratio of the masses of the atoms in the material be not too great, supposing a diatomic crystal is being considered. Steigmeier and Kudman (1966) believe such processes to be significant in materials with diamond and zinc blende lattices.

The three-phonon processes will always be present and are characteristic of the material in perfect single-crystal form. There are also collision processes involving four or more phonons which derive from the higher order anharmonicity of the interatomic forces, but these are probably not significant except at rather high temperatures. This question has been considered recently by Joshi *et al.* (1970). The remaining scattering mechanisms arise from crystal inperfections of one kind or another or from the necessarily finite size of an actual specimen. The most important is that due to the presence of point defects of various kinds. The point defects exert their influence because of their difference from the host atoms in mass, in interatomic force constants, or by causing a distortion of the surrounding crystal lattice (strain field) owing to their different size. In any otherwise perfect crystal there may be a scattering of phonons due to the presence of isotopes of different mass. Experiments on lithium fluoride and germanium single crystals containing different proportions of their isotopes have proved the importance of this factor. Foreign atoms, either deliberately added or present as impurities, will often act in all three of the ways listed above. There may also be vacant lattice sites or atoms present in interstitial positions. Other properties of a foreign atom may turn out to be significant, such as its magnetic moment, or action as a donor or acceptor in a semiconductor. Let us first of all restrict our attention to the effects of mass difference, and touch on the others subsequently.

The thermal conductivity of a crystal containing a low concentration of atoms of different mass from that of the others will be affected in two ways. Firstly phonons will be scattered by the mass difference, and secondly the phonon spectrum defined by $g(\omega)$ will be altered. The latter is usually not a very significant factor and we shall confine ourselves to noting that in the case of atoms lighter than normal, localised modes are produced whose frequencies are above the normal band of travelling-wave frequencies. The scattering of waves by defects much smaller than their wavelength is for most purposes adequately described by Rayleigh's formula which gives $\tau^{-1} \propto \omega^4$, but attention has recently become focused on what is called resonant scattering which can occur when the foreign atom is heavier than normal. The calculation of these effects is considerably more difficult, and it would not be accurate to pretend that the theory is at present in a completely satisfactory condition.

Making certain plausible assumptions (Klemens, 1955), one can write the relaxation time for scattering by mass differences

$$\tau^{-1} = \frac{\pi\Gamma\omega^2 g(\omega)}{6N} , \tag{4.24}$$

where Γ measures the strength of the scattering:

$$\Gamma = \sum_i f_i \left(1 - \frac{M_i}{M}\right)^2 ;$$

f_i is the fraction of sites occupied by atoms of mass M_i. The average mass is given by

$$M = \sum_i f_i M_i .$$

It will be seen that the effect depends on the square of the mass differences. The density of states comes in as a measure of the number of states into which the scattered phonon can go. The Rayleigh expression can be obtained by putting in the Debye form for $g(\omega)$:

$$\tau^{-1} = \frac{\Gamma\omega^4}{4\pi N v^3} . \tag{4.25}$$

This is the equation used in almost all applications of the theory.

The results in (4.24) and (4.25) are usually derived in a way which prevents resonance scattering from being considered. The simplest way of incorporating resonance is to use a classical continuum model, in which case the following expression is found:

$$\tau = \frac{w}{4\pi N}\left(\frac{\Delta M}{M}\right)^2 \frac{\omega^4}{v^3 \Pi}$$

where

$$\Pi = \left[1 - \frac{\Delta M}{M}\frac{3\omega^2}{2\omega_D^2}\left(2 + \frac{\omega}{\omega_D}\ln\frac{|\omega_D - \omega|}{\omega_D + \omega}\right)\right]^2 + \frac{9\pi}{4}\left(\frac{\Delta M}{M}\right)^2\frac{\omega^6}{\omega_D^6} \quad . \qquad (4.26)$$

In equation (4.26) Γ has been replaced by the form appropriate to a small concentration w of one foreign atom of mass $M + \Delta M$. If there are several different masses present then their effect can be obtained by summing terms like (4.26). From a plot of the term $1/\Pi$ against ω/ω_D (figure 4.11) it is seen that for heavier-than-average atoms there is a resonance effect, but that for lighter atoms the scattering effect is reduced. One curve plotted is in fact appropriate to a vacancy where $\Delta M = -M$. As well as expressions of the form given in (4.26) empirical resonant denominators have been tacked on to the Rayleigh formulation in attempts to fit experimental data. For a review of the present situation from a fundamental point of view reference should be made to recent work by Maradudin (1966) and Klein (1969).

We now turn to force constant and strain field effects. The force constant changes can be expressed in terms of changes in effective elastic constants and then incorporated in the parameter Γ. In doing this the relative change in force constant should be added to the relative change in atomic mass ($\Delta M/M$) and then the sum squared, rather than squaring first and then summing. This is because the two scattering processes are coherent and could indeed tend to cancel one another out if, for instance, a light impurity atom also brought with it a reduction in band stiffness. A similar remark applies to strain field effects.

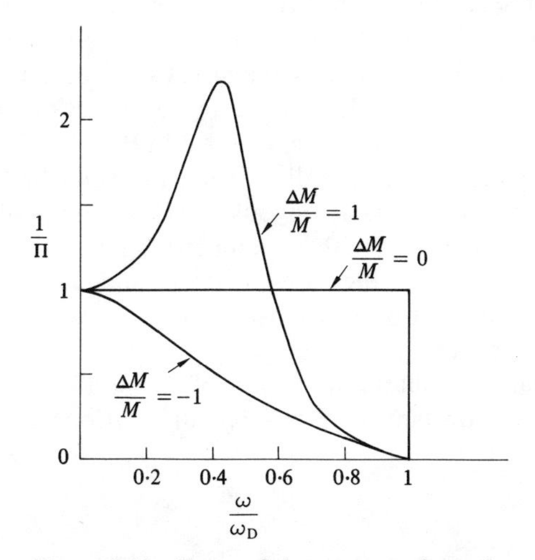

Figure 4.11. Form of the resonant denominator in equation (4.26).

Strain field scattering, which will also be encountered later in the case of scattering by dislocations, has the same physical basis as the three-phonon processes. In this latter case we have a dynamic strain field set up by the lattice vibrations; because of anharmonicity this modulates the sound velocity and causes interactions between phonons. Point defects and dislocations cause a static strain field, and once again anharmonicity results in a variation of the sound velocity in the neighbourhood of the lattice defect which scatters the phonons. Further details of the appropriate way to treat these kinds of scattering can be found in the work of Abeles (1963) and Bross (1962). Resonance effects have been treated by Krumhansl (1965).

It sometimes occurs as a result of heat treatment that impurities originally in solution in a solid are precipitated out as small particles of a different phase. In such cases we shall be concerned with scattering of lattice waves by objects larger than their wavelength. The situation is then described by a scattering cross section σ independent of frequency of the phonons so that $\tau^{-1} = N\sigma v$, where N is the number density of the precipitate particles. The cross section to be used will be some kind of average of the cross sections of the individual particles.

In some materials, particularly where there is a history of plastic deformation, scattering of phonons by dislocations plays an important role in limiting thermal conductivity. There are two possible effects here. The core of the dislocation constitutes a region of concentrated disordered crystal and should therefore scatter phonons. Geometrical arguments show that in this case $\tau^{-1} \propto \omega^3$. However, it is generally believed that the more important source of scattering is the strain field which falls off very slowly with distance from the dislocation line. The first major attempt to calculate this was made by Klemens (1955) and subsequent calculations have refined his results without generally making any essential difference. Exceptions to this are the work of Ishioka and Suzuki (1963) who considered the effects of vibrations of the dislocation line, and Brown (1967) who included resonance effects analogous to those found with point defects. Comparison of theory with experiment is not very favourable at present, particularly in the case of alkali halides where the experimentally determined dislocation resistance appears to be about a hundred times larger than theoretical predictions.

In Klemens's calculation the anharmonic effect of the strain field was specified by the Grüneisen's constant γ. For screw dislocations arranged at random

$$\tau_s^{-1} = \frac{2^{3/2} N_d b^2 \gamma^2 \omega}{27(3^{1/2} + 2^{1/2})} \tag{4.27}$$

where N_d is the number of dislocation lines cutting unit area and b is the magnitude of Burgers vector. A similar expression holds for edge dislocations:

$$\tau_{e}^{-1} = \frac{2^{3/2} N_{d} b^{2} \gamma^{2} \omega}{27(3^{1/2}+2^{1/2})} \left\{ \frac{1}{2} + \frac{\beta^{2}}{24} \left[1 + 2^{1/2} \left(\frac{v_{l}}{v_{t}} \right)^{2} \right]^{2} \right\} \tag{4.28}$$

where $\beta = (1-2\nu)/(1-\nu)$, ν is Poisson's ratio, v_l relates to longitudinal phonons and v_t to transverse electrons. In view of the approximations the numerical terms should not be treated too seriously. A similar calculation by Carruthers (1961) gives a somewhat higher scattering rate but the same frequency dependence. Analysis of experimental data suggests that this frequency dependence is in fact the correct one.

In the discussion of phonon scattering by crystal defects given above the assumption has been implicitly made that we are dealing with basically good crystalline material in which the number of defects is relatively small. Despite this, equation (4.24) has been used with considerable success in the discussion of alloys (mixed crystals) where there is a high degree of disorder in terms of distribution of chemically different atoms over the available atomic sites. The treatment of material which is really not crystalline at all is quite another question. We refer here to glasses, polymers, and amorphous substances. A theoretically sound treatment would really have to go back to scratch and consider the modes of vibration of such a disordered system, but little of practical value has yet come from this approach. Ziman (1960) has discussed the question from the more manageable standpoint of the scattering of sound waves in such a medium and arrived at some useful conclusions. At long wavelengths the reciprocal relaxation time τ^{-1} is proportional to $q^{2}\overline{(\delta v)^{2}} l_{0}/v$, where δv is the fluctuation in the sound velocity and l_0 is the scale over which these fluctuations occur. At short wavelengths τ^{-1} tends to a constant value. However, all these results are essentially qualitative in character.

The scattering mechanisms described above are applicable to all materials whether or not they conduct electricity. Those to be discussed now are only significant in the case of metals or semiconductors. Let us consider first the direct scattering of phonons by free electrons. Actually this is a rather inappropriate way of expressing it, because what actually occurs is the emission or absorption of phonons by the electrons (or holes). From the electronic viewpoint these are the same processes which cause the electrical resistance at high temperatures. The calculation of the phonon relaxation time was made by Ziman (1956) who used the effective mass approximation. The result is

$$\tau^{-1} = \frac{E_{1}^{2} m^{*2} k_{B} T}{2\pi \hbar^{4} \hat{\rho} v_{l}} \ln \frac{1+\exp(\Psi+\hbar\omega/2k_{B}T)}{1+\exp(\Psi-\hbar\omega/2k_{B}T)}, \tag{4.29}$$

where

$$\Psi = \frac{\mathscr{E}_{F}}{k_{B}T} - \frac{\hbar\omega^{2}}{8m^{*}v_{l}^{2}k_{B}T} + \frac{m^{*}v_{l}^{2}}{2k_{B}T}$$

for scattering of longitudinal phonons by electrons. In this approximation there is no provision for scattering of transverse electrons but it is usually taken as being similar in strength. In (4.29) E_1 is a constant specifying the strength of the electron–phonon interaction, called the deformation potential. The scattering rate depends on the number of electrons through the Fermi energy, $\mathscr{E}_F$.

In the case of metals or degenerate semiconductors the expression given above simplifies to

$$\tau^{-1} = \frac{E_1^2 m^{*2} \omega}{2\pi\hbar^3 \hat{\rho} v_1} . \tag{4.30}$$

This result really only applies where the Fermi surface is spherical, but it is probable that the frequency dependence is not too sensitive to Fermi surface geometry. An expression can also be derived from (4.29) for the case of semiconductors with small electron concentration but then the effect is usually negligible.

The result given in equation (4.30) requires amendment in the case of impure metals at very low temperatures, when the electron mean free path becomes shorter than the phonon wavelength. Pippard (1955) showed that the theory underlying the results given above then breaks down, the coupling between phonons and electrons becoming electromagnetic in nature. One consequence of this is that there is appreciable scattering of transverse phonons. At these very long wavelengths τ^{-1} becomes proportional to ω^2 and to the electron mean free path; the subject has been reviewed by Spector (1966).

An important change in the scattering of phonons by electrons occurs if a metal becomes superconducting. Any detailed discussion would carry us rather deeply into the phenomena and theories of superconductivity (see e.g. Lynton, 1964), but the general idea emerges in a straightforward way from the two-fluid model. According to this, below the superconducting transition temperature, T_c, an increasing fraction of the electrons 'condense' into a state in which they have no scattering interaction with phonons. The consequences of this have been worked out by Bardeen *et al.* (1959). They show that, if τ_N is the normal relaxation time for scattering of phonons by electrons and τ_{sc} is the same quantity when the metal has become a superconductor, then

$$\tau_{sc}^{-1} = g(z)\tau_N^{-1} , \tag{4.31}$$

where

$$z = \frac{\hbar\omega}{k_B T},$$

$$g(z) = \frac{1-e^{-z}}{z}(2J_1 + J_2),$$

$$J_1 = \int_{\mathcal{E}_0}^{\infty} \left(\frac{\mathcal{E}^2 + Ez}{\mathcal{E}\mathcal{E}'}\right) f_0(E) f_0(-E') \mathrm{d}E,$$

$$J_2 = \int_{-z+\mathcal{E}_0}^{-\mathcal{E}_0} \left|\frac{EE'}{\mathcal{E}\mathcal{E}'}\right| \left(1 - \frac{\mathcal{E}_0^2}{EE'}\right) f_0(E) f_0(-E') \mathrm{d}E,$$

$$E' = E + z, \qquad E = (\mathcal{E}^2 + \mathcal{E}_0^2)^{1/2},$$

$\mathcal{E}$ is the normal electron energy and $2\mathcal{E}_0$ the superconducting energy gap. At very low temperatures $\tau_{sc}^{-1} \to 0$.

In n-type semiconductors the free electrons are produced by the thermal ionisation of what are called donor atoms which therefore become positively charged. An analogous process occurs in p-type semiconductors. At low temperatures, however, the electrons are trapped by the donors which in consequence are now neutral. Neutral donors can be highly significant scatterers of phonons in many semiconductors and the theory of this has been worked out by Keyes (1961) and Griffin and Carruthers (1963) to whom reference should be made for details of the relaxation times.

The lattice thermal conductivity is appreciably modified in materials containing paramagnetic ions, particularly in the presence of applied magnetic fields. These might be either diamagnetic materials to which a small concentration of paramagnetic ions has been added, or materials in which one of the normal constituents is paramagnetic. In the former case it is probably appropriate to treat this as a scattering problem and Orbach (1962) has shown how to do this. A more rigorous treatment applicable to both cases requires consideration of the effect on the lattice vibrational spectrum $g(\omega)$ of the spin–phonon coupling. This has been the approach of Eliott and Parkinson (1967). Details will not be given but it should be noted that the effect is only large at very low temperatures. In ferromagnetic insulators another type of phonon scattering is possible at low temperatures which is due to interaction with spin waves. These are coupled vibrations of the atomic magnetic moments; the quanta of these vibrations are called magnons. An account of the physical nature of spin waves can be found in Kittel's book (1970). Joshi and Sinha (1966) have shown how to calculate this scattering and found that $\tau^{-1} \propto q^3$. Magnons will be referred to again in section 4.9 where they appear in the role of carriers of heat.

The remaining important scattering process is that due to the boundaries of the conductor, including internal boundaries if the material is not a single crystal. The consequences of this were seen by Casimir (1938) who calculated the equivalent mean free path as $\ell = 2R$ for a cylindrical specimen of radius R and $\ell = 1{\cdot}12\ L$ for a square specimen of edge length L. However, Casimir neglected all other scattering processes, having furthermore assumed perfect roughness of the surface.

If there are other scattering processes then a common approximation is to add a term v_s/ℓ to the summation of reciprocal relaxation times given in (4.21). This may not be accurate, because what we have is in fact a spatially varying N_s^q. Close to the surface N_s^q will be almost $\tilde{n}$, the equilibrium form, whilst a long way from a surface it will have the value appropriate to a bulk (actually infinite) conductor. A rigorous treatment involves solving the Boltzmann equation with this kind of boundary condition. A particular difficulty arises where the only significant bulk scattering is due to N-processes, as might occur in a very pure crystal.

The assumption of perfect roughness may also not be correct. If the surface is, on the contrary, perfectly reflecting, then the scattering effect is completely lost. In fact one expects that partial reflection might occur specified by a reflection parameter varying from one for perfect reflection to zero for perfect roughness. Furthermore this reflection parameter will presumably be a function of phonon wavelength, angle of incidence, etc. For a discussion of this point and also for corrections for finite sample length reference should be made to the paper by Berman *et al.* (1955).

The principal scattering mechanisms acting on phonons have now been reviewed. The next step must be to examine the ways in which they are used in the calculation of thermal conductivity, with particular reference to the problem posed by the N-processes.

4.6 The calculation of lattice thermal conductivity

In section 4.4 the variational and relaxation time methods of calculating lattice thermal conductivity were introduced. Mostly we shall be using the relaxation time method, but a few words on the variational method are probably in order. The advantages of the variational method are that the N-processes present no special problem and that the difficulty of finding well defined relaxation times is avoided. A major disadvantage is that it really only yields numerical results, not an algebraic expression for thermal conductivity. Its best sphere of application comes in the study of specific models, because the results obtained are likely to be genuinely due to the nature of the model and not to approximations involved in coping with N-processes and ill-defined relaxation times. In this context the variational method has been recently employed by Hamilton and Parrott (1969) in a number of investigations.

If the relaxation time method is to be employed, the natural starting point is equation (4.19) for the thermal conductivity, where the total relaxation time τ_s has been obtained from those due to individual scattering mechanisms by reciprocal addition as in equation (4.21). But this formulation does not tell us how to take account of N-processes for which we require some correction term to equations (4.19) or (4.21). There are two methods of doing this which will be described in turn. However, both methods use the Debye approximation for $g(\omega)$, and this makes it convenient to use a new notation. Let us set $z = \hbar\omega/k_B T$; then equation (4.19) becomes:

$$\lambda_{\rm ph} = \frac{k_{\rm B}}{6\pi^2}\left(\frac{k_{\rm B}T}{\hbar}\right)^3 \sum_s \frac{1}{C_s} \int_0^{\Theta_s/T} \tau_s z^4 \tilde{n}(\tilde{n}+1)\,{\rm d}z = \frac{1}{3}\sum_s C_s v_s^2 \langle\tau_s\rangle\ , \tag{4.32}$$

where C_s is the heat capacity of the s-polarised phonons, Θ_s is their Debye temperature, and $\langle\tau\rangle$ is an average defined by

$$\langle\tau\rangle = \frac{\displaystyle\int_0^{\Theta/T} \tau z^4 \tilde{n}(\tilde{n}+1)\,{\rm d}z}{\displaystyle\int_0^{\Theta/T} z^4 \tilde{n}(\tilde{n}+1)\,{\rm d}z}\ . \tag{4.33}$$

In the interests of simplicity we will assume for the present that the velocities, etc., of each polarisation are the same. If this assumption is made and the N-processes are entirely neglected in evaluating τ^{-1}, equation (4.32) is often referred to as the Debye formula for thermal conductivity.

Callaway (1959) put forward a method based on the fact that if

$$N_s^q = \tilde{n} + \boldsymbol{q}\,.\,\boldsymbol{u}\tilde{n}(\tilde{n}+1), \tag{4.34}$$

where $\boldsymbol{u}$ is a constant vector, the N-processes have no effect at all. This is because such a phonon distribution corresponds to a phonon gas in equilibrium as far as energy is concerned but moving as a whole with a velocity proportional to $\boldsymbol{u}$. Now the total momentum of all the phonons can be written as

$$\boldsymbol{P} = \hbar \sum_{q,\,s} \boldsymbol{q} N_s^q\ . \tag{4.35}$$

If N_s^q has the form given in equation (4.34) then we have

$$\boldsymbol{P} = \hbar \sum_{q,\,s} \boldsymbol{q}\boldsymbol{q}\,.\,\boldsymbol{u}\tilde{n}(\tilde{n}+1). \tag{4.36}$$

What Callaway proposed was that the N-processes tend to return the phonon system not to equilibrium, $\tilde{n}$, but to a distribution given by expression (4.34), where $\boldsymbol{u}$ is given by the total phonon momentum through equations (4.35) and (4.36), N_s^q being the actual distribution. When the consequences of this assumption are worked out, expression (4.32) is replaced by

$$\lambda_{\rm ph} = \frac{1}{3}C_v v^2 \left[\left\langle\frac{\tau_{\rm R}\tau_{\rm N}}{\tau_{\rm R}+\tau_{\rm N}}\right\rangle \left\langle\frac{1}{\tau_{\rm R}+\tau_{\rm N}}\right\rangle + \left\langle\frac{\tau_{\rm R}}{\tau_{\rm R}+\tau_{\rm N}}\right\rangle^2\right] \Bigg/ \left\langle\frac{1}{\tau_{\rm R}+\tau_{\rm N}}\right\rangle\ . \tag{4.37}$$

Here $\tau_{\rm N}$ is the relaxation time for N-processes only and $\tau_{\rm R}$ is the total relaxation time for all collisions in which momentum is not conserved. Suppose that $\tau_{\rm N} \gg \tau_{\rm R}$ so that the N-processes are relatively weak. Then the second term vanishes and we are left with

$$\lambda_{\rm ph} = \tfrac{1}{3}C_v v^2 \langle\tau_{\rm R}\rangle\ , \tag{4.38}$$

which is the same as expression (4.32) with the N-processes neglected. If on the other hand the N-processes are very strong so that $\tau_N \ll \tau_R$, we find

$$\lambda_{ph} = \tfrac{1}{3}C_v v^2 \frac{1}{\langle\tau_R^{-1}\rangle} \ . \tag{4.39}$$

This is the so-called Ziman limit which has the special property of additive resistances. Using (4.21) one can see that

$$\frac{1}{\lambda_{ph}} = \frac{3}{C_v v^2} \sum_i \langle\tau_i^{-1}\rangle \ , \tag{4.40}$$

where, of course, the N-processes are not included in the summation. Only in this situation of very strong N-processes may the individual resistances of the different scattering mechanisms be added.

The Callaway method has been very widely used in the analysis of experimental data as will be seen in chapter 5. It must be stressed that in no sense was it derived from the Boltzmann equation, but it has a very strong intuitive appeal and certainly can claim the justification of success. An alternative and in some ways simpler method has been proposed by Guyer and Krumhansl (1966). This again is based on the peculiar properties of the N-processes and is much more directly derived from the Boltzmann equation, although its physical significance is more difficult to appreciate. The thermal conductivity equation is

$$\lambda_{ph} = \tfrac{1}{3}C_v v^2 \frac{\langle\tau_R\rangle}{\langle\tau_R\rangle + \langle\tau_N\rangle}\left(\langle\tau_N\rangle + \frac{1}{\langle\tau_R^{-1}\rangle}\right) \ ; \tag{4.41}$$

in the limiting cases of very weak and very strong N-processes it leads directly to equations (4.38) and (4.39) as did the Callaway equation. Up to the present time the Guyer–Krumhansl method has not been very much used, possibly because its derivation is rather obscure. It therefore remains to be seen whether it is capable of accounting for experimental results as well as the Callaway equation.

One situation which is well handled by the Guyer–Krumhansl equation is that of boundary scattering. In applications of the Callaway method boundary scattering is incorporated by simply adding v/ℓ to the reciprocal relaxation time τ_R^{-1}. This treatment is unable to account for certain phenomena of an essentially hydrodynamic nature (Poiseuille flow) which arise when N-processes are very strong. In this situation we have a flow of phonons which is unaffected by boundary scattering except in a layer within about $\tau_N v$ of the surface. Thus, compared with a conductor with no N-processes, the scattering rate due to the boundary scattering is reduced by a fraction $v\tau_N/\ell$, giving an effective mean free path of $\ell^2/v\tau_N$. In this case additional scattering has increased the mean free path. Guyer and Krumhansl (1966) have shown how to allow for this effect and their modified equation is then

$$\lambda_{ph} = \tfrac{1}{3}C_v v^2 \frac{\langle\tau_R^B\rangle}{\langle\tau_R^B\rangle + \langle\tau_N\rangle}\left[\langle\tau_N\rangle + \frac{G(r/l)}{\langle\tau_R^{-1}\rangle}\right] \ , \tag{4.42}$$

where

$$G(r/l) = 1 - \frac{2J_1(\mathrm{i}r/l)}{(\mathrm{i}r/l)J_0(\mathrm{i}r/l)},$$

r is the radius of a cylindrical sample and

$$l = \left(\frac{\langle\tau_N\rangle v^2}{5\langle\tau_R^{-1}\rangle}\right)^{1/2}$$

τ_R is obtained by reciprocal addition of all processes within the conductor in which momentum is not conserved, and

$$\frac{1}{\tau_R^B} = \frac{1}{\tau_R} + \frac{v}{\ell} = \frac{1}{\tau_R} + \frac{v}{2r}\,.$$

The Ziman limit is now

$$\lambda_{ph} = \tfrac{1}{3}C_v v^2 \frac{G(r/l)}{\langle\tau_R^{-1}\rangle} \approx \tfrac{1}{3}C_v v^2 \times \frac{5r^2}{4v^2\langle\tau_N\rangle} \qquad (4.43)$$

in agreement with the qualitative argument given earlier. At present this situation has been observed only once, in solid helium.

In the case of three-phonon processes the only relaxation times available are of the single-mode type given in table 4.1. Fortunately there are reasons why their use may not be too inaccurate. In most cases, and particularly at low temperatures, it is the low frequency phonons which contribute most to the thermal conductivity. Now, the most important scattering processes for these phonons will be those of class I and class II, where for N-processes $\tau^{-1} \propto \omega$ and where a large proportion of the collisions are with phonons of much higher frequency. Since these phonons are much closer to equilibrium than those of low frequency, it will be seen that for these collisions the assumptions of the single-mode relaxation time calculations are likely to be almost correct. Except where it is desired to draw attention to the distinction, the asterisk denoting single-mode relaxation time will be dropped in future calculations.

The Callaway method has been considerably elaborated by a number of authors. Parrott (1971) has made the alterations necessary to account separately for the contributions of the different polarisations and Holland (1963) has shown how to incorporate dispersion of the normal mode frequencies. Finally, if one refers to equation (4.23a) it will be seen that for U-processes involving the reciprocal lattice vector $\boldsymbol{K}_i$, the quantity $\boldsymbol{q} - \boldsymbol{K}_i$ is conserved in just the same way as $\boldsymbol{q}$ is conserved for N-processes. Hamilton (1973) has allowed for this by adding further terms to Callaway's equation.

The application of equations (4.37) or (4.41) to particular cases will not be proceeded with now, but delayed until chapter 5 in the context of the prediction of thermal conductivity and the analysis of experimental data. The framework required for this has, however, been set up in this present section. It is now necessary to consider the electronic thermal conductivity.

4.7 Heat conduction by electrons

The heat current carried by the conduction electrons in a metal or semiconductor can be readily written down as

$$U = 2\sum_{k} \mathscr{E}(k) u(k) f(k), \tag{4.44}$$

which is very similar in form to the corresponding result for phonons, equation (4.10). Here $u(k)$ is the velocity of an electron with wave vector k and $f(k)$ is the distribution function for electrons. In equilibrium $f(k) = f_0$, the Fermi function, and then U vanishes on account of the symmetry of $\mathscr{E}(k)$ and $u(k)$ as, of course, it must. The factor two in equation (4.44) allows for the two possible spin states of the electrons. Now $u(k)$ is proportional to the gradient of $\mathscr{E}$ with respect to k, $\nabla_k \mathscr{E}$, so if the form of $\mathscr{E}(k)$ is known then what remains is to determine $f(k)$. As in the phonon case this is normally done by solving a Boltzmann equation for electrons. An extra complication arises because electrons are affected by an electric field as well as by a temperature gradient. The problem of deriving the electron Boltzmann equation containing a temperature gradient was solved by McIrvine (1959) and Greenwood (1962). For a review of this whole topic reference should be made to Dresden (1961).

The fact that $f(k)$ is a function of electric field as well as temperature gradient means that some condition is required to fix the value of the electric field. It was noted in section 1.2 that for the thermal conductivity of an electrical conductor to be well defined it was necessary to specify that there was no electric current flowing [equation (1.3)]. This in effect fixes the electric field as a function of temperature gradient, and is the starting point of a calculation of the Seebeck coefficient. The expressions for electronic thermal conductivity therefore tend to be somewhat long-winded.

The solution of the Boltzmann equation follows similar lines to the phonon case. However, since the conditions for the use of a relaxation time are usually satisfied, the electron case is rather simpler. Only for the situation of scattering of electrons by phonons at low temperatures does any difficulty arise, necessitating the use of variational methods. As a rather important point emerges from this situation, it is desirable to consider it in a little detail, from a physical point of view.

Let us firstly consider the electron distribution in k-space in equilibrium. Figure 4.12a shows the form of $f(k) = f_0$ for a one-dimensional section of k-space, rather than as a function of $\mathscr{E}$ as was shown in figure 4.8. The value of f_0 when $k = k_F$ is, of course, one-half; let us label by k_1 and k_2 the values of k when f_0 is 0·9 and 0·1. In figure 4.12b we show the corresponding two-dimensional section with circular contours for k_1, k_F and k_2. The overall symmetry corresponds to the fact that there is neither electric nor heat current present.

Now suppose that we apply an electric field in the $-k_x$ direction. This will tend to move the electrons in the $+k_x$ direction and the one-dimensional diagram corresponding to figure 4.12a is shown in figure 4.13a. The two-dimensional case is shown in figure 4.13b. What is shown in these figures is a result of the balance between the electric field tending to displace the distribution in the $+k_x$ direction and the scattering which attempts to restore equilibrium. After taking account of the negative charge on the electron it will be seen that there is a net electric current in the direction of the electric field.

Scattering by phonons results in a discontinuous jump by the electrons from one state $\boldsymbol{k}$ to another $\boldsymbol{k}'$, the phonon wave vector being of magnitude $|\boldsymbol{k}-\boldsymbol{k}'|$. Consider two different possible scattering events labelled A and B on diagrams 4.13a and 4.13b. In process A a very large change in $\boldsymbol{k}$ occurs which tends to reduce greatly the electric current, but it requires

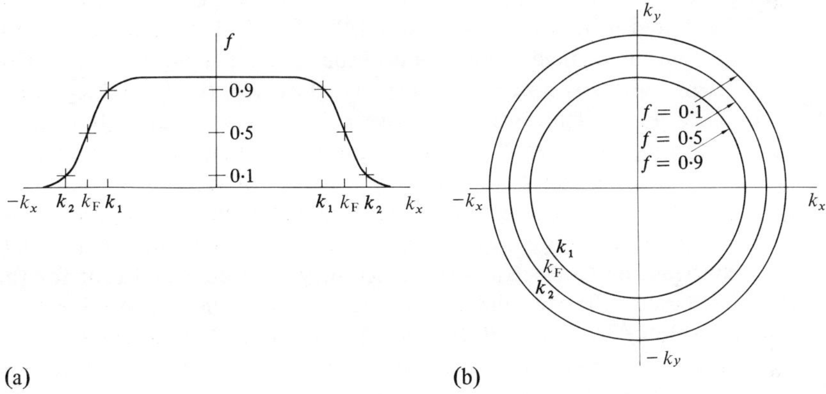

Figure 4.12. Electron distribution in equilibrium: (a) one-dimensional section, (b) two-dimensional contours.

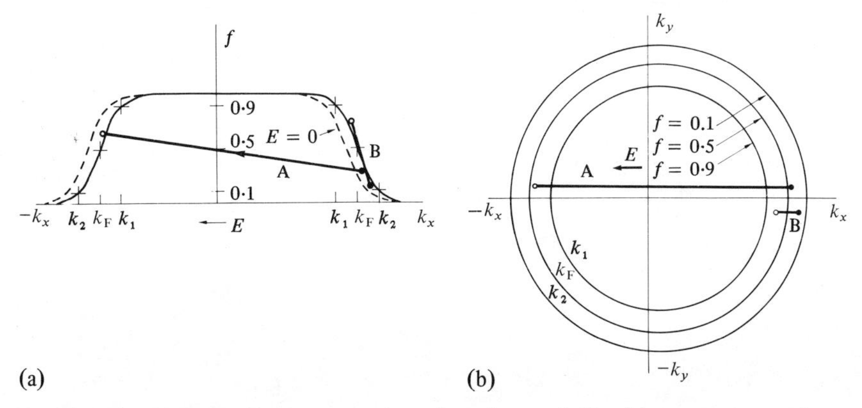

Figure 4.13. Electron distribution with applied electric field: (a) one-dimensional section, (b) two-dimensional contours.

the collaboration of a high frequency phonon. In process B there is a very small change in $\boldsymbol{k}$, so such events contribute little to the electrical resistance, but only low energy phonons are needed. At room temperature there are plenty of phonons of high and low energies, but at low temperatures the high-energy phonons tend to disappear, as expressed in the Planck function [equation (4.4)]. Thus process A tends to be suppressed at low temperatures. In fact the collisions are not of two types only, but occur over the whole spectrum of $\boldsymbol{k}-\boldsymbol{k}'$ values. It will be clear that not only will there be fewer electron–phonon collisions at low temperatures, but those that do occur will be less effective in giving rise to electrical resistance. In fact, if there is very little scattering of electrons by crystalline imperfections, the electrical conductivity rises very rapidly at low temperatures.

We next consider the situation where there is a temperature gradient present but no electric current. The resulting distribution function is illustrated in figures 4.14a and 4.14b. Firstly it should be noticed that the k_{F} contour coincides with its position in thermal equilibrium. This is because there is no electric current. However, the k_1 and k_2 contours are displaced in such a way as to give a finite heat current. Here the scattering processes A and B become of comparable importance. Processes of type A are called 'horizontal', processes of type B 'vertical'. Horizontal processes give a change of momentum and energy, while vertical processes change only energy appreciably. Thus, scattering which is ineffective in contributing to electrical resistance may be quite significant for thermal resistance. Clearly, under such circumstances the use of relaxation times is hazardous. Also it will be obvious that thermal conductivity will tend to increase less than electrical conductivity as the temperature decreases, so that here we have the reason for the breakdown in the Wiedemann–Franz law at low temperatures. This point is discussed further by Jones (1956) and Ziman (1960).

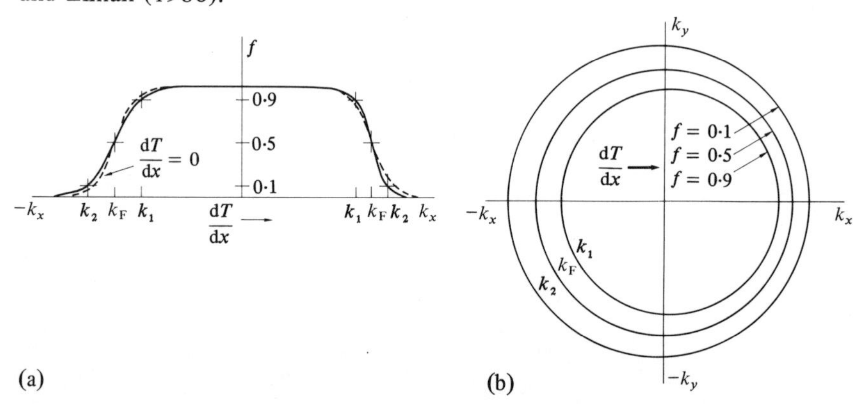

Figure 4.14. Electron distribution with temperature gradient but no electric current: (a) one-dimensional section, (b) two-dimensional contours.

To show the reason for the validity of the Wiedemann–Franz law in most situations we will now consider the equations for thermal and electrical conductivity in the relaxation-time approximation (Wilson, 1953). The electronic thermal conductivity λ_e is given by

$$\lambda_e = \frac{1}{3k_B T^2}\left(M_2 - \frac{M_1^2}{M_0}\right), \tag{4.45}$$

where

$$M_i = \int u^2 \tau \mathcal{E}^i f_0(1-f_0) N(\mathcal{E})\, d\mathcal{E}, \tag{4.45a}$$

and the electrical conductivity by

$$\sigma = \frac{e^2 M_0}{3k_B T}. \tag{4.46}$$

Now the Lorenz number is defined by $\mathcal{L} = \lambda/\sigma T$; so, using (4.45) and (4.46), we obtain

$$\mathcal{L} = \frac{1}{e^2 T^2}\left[\frac{M_2}{M_0} - \left(\frac{M_1}{M_0}\right)^2\right]. \tag{4.47}$$

This expression has the form of a weighted mean square deviation of energy. If we define an average of Λ by

$$\overline{\Lambda} = \frac{\int u^2 \tau \Lambda f_0(1-f_0) N(\mathcal{E})\, d\mathcal{E}}{\int u^2 \tau f_0(1-f_0) N(\mathcal{E})\, d\mathcal{E}}, \tag{4.47a}$$

then

$$\mathcal{L} = \frac{1}{e^2 T^2}\left[\overline{\mathcal{E}^2} - \overline{\mathcal{E}}^2\right].$$

One might therefore expect that $\mathcal{L} = \alpha(k_B/e)^2$ since such a squared deviation of energy would normally take the form $\alpha(k_B T)^2$, where α is a number of order one. Exact calculations reveal that this is indeed the case.

The equation (4.45), (4.46), and (4.47) are appropriate not only for metals but also for semiconductors. The difference lies in the form of $f_0(1-f_0)$. In the case of a metal $\mathcal{E}_F$ is positive and larger than $k_B T$, and f_0 takes the form shown in figure 4.8 and $f_0(1-f_0)$ that shown in figure 4.9. Thus the presence of the $f_0(1-f_0)$ factor means that the values of the integrals depend on the form of the integrands in the neighbourhood of the Fermi level, $\mathcal{E}_F$. For semiconductors $\mathcal{E}_F \ll 0$, and the Fermi function is approximated by equation (4.12) and $1-f_0 \approx 1$. Then

$$\overline{\Lambda} = \frac{\int u^2 \tau \Lambda \exp(-\mathcal{E}/k_B T) N(\mathcal{E})\, d\mathcal{E}}{\int u^2 \tau \exp(-\mathcal{E}/k_B T) N(\mathcal{E})\, d\mathcal{E}}$$

and it turns out that $\mathcal{L}$ is significantly dependent on the way in which τ varies as a function of energy. This is not the case for metals where a straightforward calculation, originally due to Sommerfeld, gives

$$\mathcal{L} = \mathcal{L}_0 = \frac{\pi^2}{3}\left(\frac{k_B}{e}\right)^2 \approx 2{\cdot}45 \times 10^{-8}\ \mathrm{V^2\,K^{-2}}, \tag{4.48}$$

a result in good agreement with experiment for many metals. Note that $\mathcal{L}$ is independent of the density of states function $N(\mathcal{E})$.

Let us review the limitations of the result expressed in equation (4.48). Firstly, this equation will only apply where the lattice thermal conductivity is either quite negligible or can be accurately subtracted from the measured thermal conductivity to leave only the electronic part. Secondly, it will not be true unless $\mathcal{E}_F \gg k_B T$; for normal metals this criterion is quite well satisfied. Thirdly, it depends on the validity of the relaxation time approximation. This is correct for scattering of electrons by lattice imperfections, but, as was shown earlier in this section, may be quite misleading for the phonon scattering at low temperatures.

The most important electron scattering mechanism at moderate and high temperatures is always that due to the lattice vibrations. The reciprocal relaxation time has the form

$$\frac{1}{\tau} = \frac{E_1^2 k_B T \omega_D^4}{32\pi\hat{\rho} v_l^6 (2m^*)^{1/2} \mathcal{E}^{3/2}} \tag{4.49a}$$

for metals, and

$$\frac{1}{\tau} = \frac{E_1^2 (2m^*)^{3/2} k_B T \mathcal{E}^{1/2}}{2\pi\hbar^4 \hat{\rho} v_l^2} \tag{4.49b}$$

for semiconductors and semimetals. In these equations, E_1 is the constant used in the discussion of the scattering of phonons by electrons (deformation potential) in section 4.5, ω_D is the maximum allowed phonon frequency in the Debye approximation as discussed in section 4.2 and $\hat{\rho}$ is the density. These results only apply to a simple effective mass model but the important linear dependence on temperature is generally valid and distinguishes this kind of scattering from all others.

The other significant scattering arises from various static crystalline imperfections. These may be point imperfections such as impurity atoms or vacancies, or linear imperfections such as dislocations. A very high concentration of a particular impurity atom will correspond to a substitutional alloy and will give a great deal of scattering. However, the situation here is likely to be complicated by concurrent changes in Fermi energy, effective mass, etc. Relaxation times can be worked out for all these scattering processes (Ziman, 1960), but the most important factor really is the fact that all these relaxation times are independent of temperature.

One scattering mechanism which might appear to be significant is electron–electron scattering. An account of this is given by Jones (1956), where it is shown to be in fact rather weak due to screening effects. In the case of electrical conductivity the effect is exceedingly small because electron momentum, and therefore current, is conserved in such collisions, but even in thermal conductivity it is likely to be very difficult to detect except by extremely accurate measurements. The situation is different in the transition metals where, owing to the complexity of the Fermi surface, electron–electron scattering can contribute to both electrical and thermal resistivity. Herring (1967) has shown that in such cases the Lorenz number can be anomalously low.

All the discussion in this section has been in terms of a very simple model of a metal or semiconductor. For metals a spherical Fermi surface is assumed, whilst for semiconductors only the most straightforward isotropic effective-mass approximation has been considered. No account has been taken of the possible presence of holes in either metals or semiconductors. If it should be necessary to consider a metal or semiconductor where only holes are significant then all that is required is a simple change of sign of electric charge on the carrier and the ideas presented above will suffice. If, on the other hand, both electrons and holes are present, new phenomena appear which must now be discussed.

Let us consider an intrinsic semiconductor. In such material there exist equal numbers of electrons and holes produced by the thermal excitation of electrons from the valence to the conduction band. The number of electrons, N, and the number of holes, P, are each given by

$$P = N = (N_c N_v)^{1/2} \exp(-\mathcal{E}_g/2k_B T), \tag{4.50}$$

where N_c and N_v are quantities characteristic of the conduction and valence bands, and $\mathcal{E}_g$ is the energy gap. Thus, if there is a temperature gradient, the values of P and N will be greater in the hotter region and lower in the cooler region. Consequently there will be a flow of electrons and holes down the temperature gradient followed by recombination with the release of energy roughly equal to $\mathcal{E}_g$. This flow together with energy release provides an additional contribution to the heat current. Therefore in an intrinsic semiconductor we expect the thermal conductivity due to carriers to consist of three terms: (i) a normal term like equation (4.45) for electrons, (ii) a similar term for holes, and (iii) a term describing the transport of activation energy $\mathcal{E}_g$ by electron–hole pairs.

The condition of zero electric current can still hold since the electrons and holes carry opposite charge. In a single-carrier conductor this is achieved by the more energetic carriers flowing down the temperature gradient and the less energetic up the gradient, as can be seen from figures 4.14a and 4.14b.

The argument presented above can be readily extended to the case where the number of electrons is not equal to the number of holes and neither is negligible. Rather more surprisingly, there is an analogous effect in semimetals where there is a small overlap of the valence and conduction bands; this can be regarded as a negative energy gap.

4.8 The calculation of electronic thermal conductivity

The calculation of the electronic thermal conductivity of metals is made much simpler by the form of the Fermi function, or more precisely by the form of $f_0(1-f_0)$. This has the effect that only those electrons with energies near to $\mathcal{E}_F$ need be considered, so that the energy dependence of the relaxation times, and of $N(\mathcal{E})$, is irrelevant. Let us write τ_{ph} and τ_d for the relaxation times due to scattering by phonons and by crystalline imperfections. Then, although the total relaxation time must be obtained by reciprocal addition, since only the value of τ at $\mathcal{E}_F$ is needed this can be divided up again in terms of thermal resistivities $W = 1/\lambda_e$ thus

$$W = W_0 + W_i , \tag{4.51}$$

where W_0 is due to the defects and W_i due to phonon scattering. The suffix i stands for ideal (or intrinsic), since W_i is the ideal (or intrinsic) resistance possessed by a perfect metal free of all defects. In fact W_0 can also be broken down into terms attributable to particular lattice imperfections. This additivity of thermal resistances is a form of Matthiessen's rule.

The electrical resistivity of a metal can be divided in a similar way

$$\rho = \rho_0 + \rho_i , \tag{4.52}$$

where ρ_0 is the residual resistivity and ρ_i the 'ideal' resistivity. The relation between W_0 and ρ_0 is always given by the Sommerfeld value of the Lorenz number:

$$W_0 = \frac{\rho_0}{\mathcal{L}_0 T} = \frac{3e^2\rho_0}{\pi^2 k_B^2 T} . \tag{4.53}$$

Direct evaluation yields

$$\frac{1}{W_0} = \tfrac{1}{9}\pi^2 k_B^2 T u_F^2 \tau_{dF} N(\mathcal{E}_F) \tag{4.54}$$

where u_F and τ_{dF} are the velocity and defect relaxation time at the Fermi energy. Now the electronic heat capacity can be shown to be

$$C_v = \tfrac{1}{3}\pi^2 k_B^2 T N(\mathcal{E}_F) \tag{4.55}$$

whilst the mean free path ℓ_{dF} of the electrons is $u_F\tau_{dF}$, whence

$$\frac{1}{W_0} = \tfrac{1}{3} C_v u_F \ell_{dF} \tag{4.56}$$

which is the usual kinetic theory form. It will be seen from (4.54) that, since τ_{dF} is independent of temperature, the thermal resistance due to defects is inversely proportional to temperature.

The theoretical calculation of ℓ_{dF} to any great accuracy is not possible. In the case of chemical impurities the scattering is approximately proportional to the square of the difference in valency between the impurity atom and the host atoms. Approximate calculations of the scattering due to vacancies, dislocations, etc., also have been made.

The discussion of phonon scattering of electrons given in section 4.7, will have made it clear that the calculation of W_i will be quite difficult except at moderate and high temperatures. Similarly, the Wiedemann–Franz law will not hold between W_i and ρ_i except at these temperatures. One complication not mentioned earlier is that electron–phonon umklapp processes may be important. The simplest form of electron–phonon interaction has

$$\boldsymbol{k}'-\boldsymbol{k} = \pm\boldsymbol{q},$$

as suggested in the discussion in section 4.7. This equation is similar to the three-phonon interaction equations discussed in section 4.5, and it can be shown that it requires generalising, in the same way as equations (4.23), to

$$\boldsymbol{k}'-\boldsymbol{k} = \pm\boldsymbol{q}+\boldsymbol{K}_i\,, \tag{4.57}$$

where $\boldsymbol{K}_i$ is a reciprocal lattice translation vector. Again, if $K_i = 0$ we say that the process is normal (N-process); otherwise one speaks of umklapp processes (U-processes). The important thing is that U-processes are (relative to N-processes) much more likely at high than at low temperatures.

An expression for the ideal thermal resistivity in which U-processes are neglected has been worked out by Wilson (1953). This is

$$W_i = \frac{12\rho_\Theta}{\pi^2 T}\left(\frac{e}{k_B}\right)^2\left\{\left[\left(\frac{T}{\Theta_D}\right)^5+\frac{3}{\pi^2}\left(\frac{k_F}{q_D}\right)^2\left(\frac{T}{\Theta_D}\right)^3\right]\mathscr{J}_5\left(\frac{\Theta_D}{T}\right)-\frac{1}{2\pi^2}\left(\frac{T}{\Theta_D}\right)^5\mathscr{J}_7\left(\frac{\Theta_D}{T}\right)\right\}, \tag{4.58}$$

where $q_D = \omega_D/v_l$ and $\Theta_D = \hbar\omega_D/k_B$. The functions $\mathscr{J}_n(x)$ are defined by

$$\mathscr{J}_n(x) = \int_0^x \frac{z^n \mathrm{d}z}{(\mathrm{e}^z-1)(1-\mathrm{e}^{-z})} \tag{4.59}$$

and are discussed in Wilson's book. The expression for the thermal resistivity is to be compared with that for the electrical resistivity

$$\rho_i = 4\rho_\Theta\left(\frac{T}{\Theta_D}\right)^5\mathscr{J}_5\left(\frac{\Theta_D}{T}\right), \tag{4.60}$$

from which it will be seen that the first term in equation (4.58) is due to 'horizontal scattering' and therefore is related to (4.60) by the normal

Lorenz number, whilst the second term is due to 'vertical scattering' and does not relate to anything in (4.60). The third term in equation (4.58) allows for the fact that vertical and horizontal scattering are not really independent.

Now the functions $\mathcal{J}_n(x)$ have the following particular properties:
(i) when x is small

$$\mathcal{J}_n(x) = \frac{x^{n-1}}{n-1},$$

(ii) when $x \to \infty$

$$\mathcal{J}_n(x) = n!\,\zeta(n),$$

where $\zeta(n)$ is the Riemann ζ-function. The limit when x is small corresponds to high temperatures and substitution in (4.60) shows that under these circumstances

$$\rho_\mathrm{i} = \rho_\Theta \left(\frac{T}{\Theta_\mathrm{D}}\right), \tag{4.61}$$

so ρ_Θ is the electrical resistivity at the Debye temperature. In the same temperature range

$$W_\mathrm{i} = \frac{3}{\pi^2}\frac{\rho_\Theta}{\Theta_\mathrm{D}}\left(\frac{e}{k_\mathrm{B}}\right)^2 = \frac{3}{\pi^2}\left(\frac{e}{k_\mathrm{B}}\right)^2\frac{\rho_\mathrm{i}}{T}, \tag{4.62}$$

so the Lorenz number has its normal value.

At very low temperatures ρ_i decreases as T^5, whereas from (4.58) we find

$$W_\mathrm{i} = \frac{36 \times 5!\,\zeta(5)\rho_\Theta T^2}{\pi^4 \Theta_\mathrm{D}^3}\left(\frac{k_\mathrm{F}}{q_\mathrm{D}}\right)^2\left(\frac{e}{k_\mathrm{B}}\right)^2, \tag{4.63}$$

so that the temperature dependence is as T^2. Clearly the Wiedemann–Franz law completely breaks down under these circumstances. However, when W_i and ρ_i become much less than W_0 and ρ_0 the latter will dominate the thermal and electrical resistivities and the measured quantities will be found to satisfy the Wiedemann–Franz law once more. In figure 4.15 are shown theoretical curves of thermal conductivity against temperature for varying amounts of defect resistance, and in figure 4.16 are shown corresponding effective Lorenz numbers.

A very important feature of figure 4.15 is the fact that it shows a minimum in the thermal conductivity which is never observed experimentally. This is believed to be because of the neglect of the umklapp processes in the above expressions. Because there is a minimum phonon energy for U-processes these are very weak at low temperatures. In figure 4.15 the thermal conductivity has been normalised by its high temperature value, and there the effect of including U-processes would be to raise the low temperature 'ideal' thermal conductivity and to tend to eliminate the minimum. Calculations by Ziman (1954) gave results in which the minimum almost completely disappeared and which were in good agreement with experiment.

A drastic change occurs in the electronic thermal conductivity in those metals which become superconducting. According to the two-fluid model the superconducting electrons have condensed into a state in which they cannot conduct heat; furthermore the fraction of electrons in these states increases steadily as the temperature falls. Thus the electronic thermal conductivity decreases steadily in a manner calculated by Bardeen *et al.* (1959). An exact result can be obtained for the situation where resistance in the normal state is completely of the residual type. Then, if λ_{en} is the normal electronic thermal conductivity and λ_{esc} is the same quantity in the

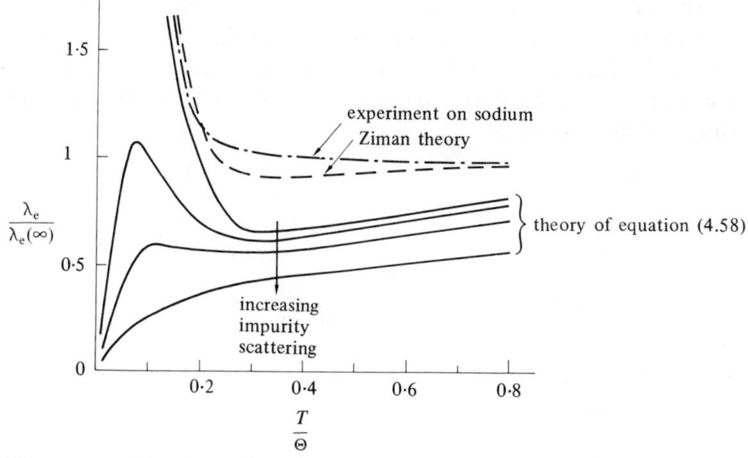

Figure 4.15. Variation of normalised electronic thermal conductivity with temperature and impurity content.

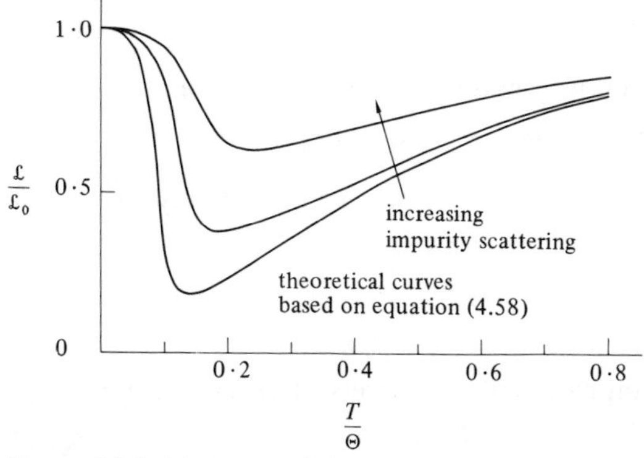

Figure 4.16. Variation of theoretical Lorenz number with temperature and impurity content.

superconducting state, we have

$$\frac{\lambda_{\text{esc}}}{\lambda_{\text{en}}} = \frac{2F_1(-y) + 2y\ln(1+e^{-y}) + y^2(1+e^{y})}{2F_1(0)} \tag{4.64}$$

where $y = \mathscr{E}_0/k_B T$ and

$$F_j(x) = \int_0^\infty \frac{x^j \mathrm{d}x}{e^{x-z}+1}\,. \tag{4.65}$$

This equation in fact means that for all superconductors $\lambda_{\text{esc}}/\lambda_{\text{en}}$ is a universal function of T/T_c, since the critical temperature T_c is determined by the energy gap, $2\mathscr{E}_0$. For the case where the resistance is partly due to phonon scattering an equation for $\lambda_{\text{esc}}/\lambda_{\text{en}}$ in integral form has been derived by Kadanoff and Martin (1961).

The calculation of the electronic thermal conductivity of simple semiconductors is a quite straightforward application of equation (4.45) whilst the Lorenz number is similarly easily derived from equation (4.47). We suppose that we have bands of standard form and that the relaxation time can be written as

$$\tau = \mathscr{D}\mathscr{E}^z, \tag{4.66}$$

where $\mathscr{D}$ is a factor accounting for any temperature dependence of τ, and z is a numerical exponent. Then the integrals

$$M_i = \mathscr{A}(i+z+\tfrac{3}{2})(k_B T)^{i+z+5/2} F_{i+z+1/2}(\mathscr{E}_F/k_B T), \tag{4.67}$$

where

$$\mathscr{A} = \frac{2^{3/2} m^{*1/2} \mathscr{D}}{\pi^2 \hbar^3}$$

and the functions $F_{i+z+1/2}$ were given by equation (4.65). Since we have already considered the case of metals, we are mainly interested now in the case where $\mathscr{E}_F/k_B T \ll 0$. In such circumstances

$$F_j(x) = \Gamma(j+1)\,e^x,$$

where $\Gamma(y)$ is the gamma function. Then our integral becomes

$$M_i = \mathscr{A}\Gamma(i+z+\tfrac{5}{2})(k_B T)^{i+z+5/2}\exp(\mathscr{E}_F/k_B T).$$

Now $\Gamma(y) = (y-1)\Gamma(y-1)$, so if we turn to equation (4.47) for the Lorenz number we find

$$\mathscr{L} = (z+\tfrac{5}{2})\left(\frac{k_B}{e}\right)^2. \tag{4.68}$$

The most important cases are (i) lattice scattering where $z = -\frac{1}{2}$ so

$$\mathscr{L} = 2\left(\frac{k_B}{e}\right)^2,$$

and (ii) ionized impurity scattering where $z = \frac{3}{2}$ so

$$\mathcal{L} = 4\left(\frac{k_B}{e}\right)^2.$$

A similar evaluation in the degenerate limit gives the standard result that $\mathcal{L} = \pi^2 k_B^2/3e^2$. However, it frequently occurs that conductors which are neither nondegenerate nor fully degenerate have to be considered. In this case numerical evaluation of the integrals $F_j(x)$ is necessary; a convenient tabulation is provided by Madelung (1957). Plots of the variations of $\mathcal{L}$ with reduced Fermi energy are shown in figure 4.17 for a number of values of z. Although all the above has been presented in terms of electrons, exactly the same results apply to holes.

The discussion of electronic thermal conductivity of semiconductors in terms of relaxation times is subject to the same limitations as occur in the case of metals, that is the scattering mechanism must be elastic. This assumption will not hold when there is scattering of electrons by optical phonons, but Korenblit and Sherstobitov (1968) have shown how to treat this problem in the case of degenerate semiconductors by using a complex relaxation time. They were able to relate the inelasticity of the scattering to a reduction in the Lorenz number below the normally expected value.

When both electrons and holes are present in comparable numbers the transport of activation energy mentioned in section 4.7 must be allowed for. The total electrical conductivity σ_t is simply the sum of the separate electron and hole conductivities.

$$\sigma_t = \sigma_e + \sigma_h. \tag{4.69}$$

We can write the total (electron + hole) thermal conductivity λ_t using a Lorenz number, as

$$\lambda_t = \mathcal{L}_t T \sigma_t$$

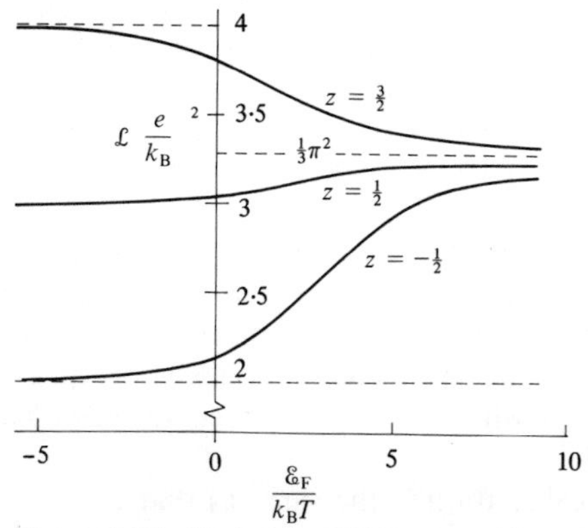

Figure 4.17. Variation of Lorenz number with reduced Fermi level for different values of scattering parameter z computed using equation (5.9).

where $\mathcal{L}_t$ is given by

$$\mathcal{L}_t = \left(\frac{k_B}{e}\right)^2 \frac{\sigma_e \sigma_h}{\sigma_t^2}\left(\frac{\mathscr{E}_g}{k_B T} + \alpha_e + \alpha_h\right)^2 + \frac{\mathcal{L}_e \sigma_e}{\sigma_t} + \frac{\mathcal{L}_h \sigma_h}{\sigma_t}\ ; \tag{4.70}$$

$\mathcal{L}_e$, $\mathcal{L}_h$ are the separate electron and hole Lorenz numbers, and α_e and α_h are given by

$$\alpha = z + \tfrac{5}{2}, \tag{4.71}$$

where appropriate values of z for electrons and holes are used. This result only applies to nondegenerate conductors; a more complicated expression is necessary where any degree of degeneracy exists. It will be seen from (4.70) that where either $\sigma_e \ll \sigma_h$ or $\sigma_h \ll \sigma_e$; the first term becomes very small, only the term corresponding to the majority carriers contributes appreciably.

The largest Lorenz number occurs when $\sigma_e \approx \sigma_h$. The energy gap of silicon, for instance, is about 45 $k_B T$, so, if $z = -\frac{1}{2}$ for lattice scattering, we have

$$\mathcal{L}_t \approx [\tfrac{1}{2}(49)^2 + 2]\left(\frac{k_B}{e}\right)^2 \approx 1200\left(\frac{k_B}{e}\right)^2.$$

This is about 600 times the normal single-carrier case. Even for bismuth telluride $\mathscr{E}_g/k_B T \approx 5$, giving $\mathcal{L}_t \approx 40(k_B/e)^2$, which is 20 times the single carrier value. Thus, even where the electrical conductivity is quite low, there may be an appreciable thermal conductivity due to electrons and holes in an intrinsic semiconductor.

4.9 Heat conduction due to other mechanisms

Heat conduction by lattice vibrations and that by electrons and holes constitute the most important mechanisms in nearly all substances at nearly all temperatures. Certainly they have attracted most of the theoretical interest, but there are other possibilities, for example heat conduction by photons or magnons.

Every material contains within it thermal radiation corresponding to its mean temperature. In equilibrium the radiation is of the 'blackbody' kind, which was so important in the discovery of quantum effects. If the conductor is transparent for a particular range of wavelength then the radiation passes through undisturbed, but if the radiation is absorbed then the photons diffuse through the material in a manner analogous to that exhibited by the lattice vibrational phonons. It is quite straightforward to calculate the corresponding thermal conductivity, as was first done by Genzel (1953). The calculation was extended to uniaxial crystals by Devyatkova *et al.* (1959). This whole subject has recently been discussed by Men' and Sergeev (1973).

The starting point of the calculation is the same as that expressed by equation (4.1) and expanded in (4.19). However the mean free path is replaced by the reciprocal of the optical absorption coefficient $\alpha(\omega)$.

Thus

$$\lambda_r = \frac{1}{3}\int_0^{\infty} \frac{C_r(\omega)v_g}{\alpha(\omega)}\,d\omega \tag{4.72}$$

where the suffix r refers to 'radiation', $C_r(\omega)d\omega$ is the heat capacity of blackbody radiation in the frequency range ω to $\omega + d\omega$, and v_g is the group velocity of the radiation, which in a solid is not the same as the free space velocity of light. The evaluation of equation (4.72) leads to

$$\lambda_r = \tfrac{16}{3}\sigma_R n^2 T^3 \langle \alpha^{-1} \rangle \tag{4.73}$$

where σ_R is the Stefan–Boltzmann constant, n is the refractive index and $\langle \alpha^{-1} \rangle$ is an average of the kind defined in equation (4.33) with the upper limit of integration set at infinity. Normally $\alpha(\omega)$ is obtained experimentally, and it is really only necessary to know it in the neighbourhood of the maximum of the function $\tilde{n}(\tilde{n} + 1)$.

The presence of the radiative heat conduction mechanism has been established in a number of semiconductors where the lattice thermal conductivity is rather low and the energy gap fairly high. Examples are tellurium, selenium, and germanium–silicon alloys; this last example will be investigated further in the next chapter.

It has been realised for some time that magnons, or quantised spin waves, might contribute to thermal conductivity in ferromagnetic insulators. The relation between frequency ω and wave number k is given by

$$\hbar\omega = Dk^2 + g\mu_B B, \tag{4.74}$$

where D is a constant of proportionality and the second term gives the effect of a magnetic field B, g is the spectroscopic splitting factor, and μ_B the Bohr magneton. The magnons follow a Planck type of distribution function [equation (4.4)] rather like the phonons. As a result the thermal conductivity due to magnons is given by

$$\lambda_m = \frac{k_B(k_B T)^2}{6\pi^2 D\hbar}\int_0^{x_{max}} \frac{(x+\delta)^2 \ell e^{x+\delta}}{(e^{x+\delta}-1)^2}\,dx \tag{4.75}$$

where $x = Dk^2 k_B T$, x_{max} is the maximum value of x, $\delta = g\mu_B B/k_B T$, and ℓ is the mean free path for magnons. The main scattering processes for magnons are collisions with phonons, defects, and boundaries. Since magnon thermal conductivity is most noticeable at low temperatures, boundary scattering is usually dominant. Thus, in zero magnetic field

$$\lambda_m = \frac{k_B(k_B T)^2 \ell_b}{6\pi^2 D\hbar}\int_0^{\infty} \frac{x^3 e^x}{(e^x-1)^2}\,dx \tag{4.76}$$

where ℓ_b is the boundary scattering mean free path. On comparing this with lattice thermal conductivity which is proportional to T^3 it will be seen why magnon conductivity is most easily observed at low temperatures.

If the magnetic field is not zero, the excitation of the magnons is much reduced by the effect of the δ term in (4.75). One method of establishing the presence of magnon conductivity is therefore to examine the effect of an applied magnetic field. The results expressed in equations (4.75) and (4.76) were obtained by Sato (1955) and Callaway and Boyd (1964). The latter authors also considered the likely effects of N- and U-type scattering.

Large magnon contributions have been found experimentally in materials like ferromagnetic EuS (McCollum *et al.,* 1964) and yttrium iron garnet (YIG) (Douglass, 1963). In the latter case 70% of the heat conduction was by magnons at 0·5 K, a result obtained by quenching the magnon conduction by an applied magnetic field. All the experimental observations to date have been at extremely low temperatures.

This completes the list of observed thermal conductivity mechanisms. There are other possible energy transport mechanisms, such as the exciton conduction believed at one time (Devyatkova, 1957) to occur in lead telluride; later measurements showed that it was unnecessary to invoke such a mechanism. However, it would be premature to assume that no other significant heat conduction processes exist in solids.

References

Abeles, B., 1963, *Phys. Rev.,* **131,** 1906.
Bardeen, J., Rickaysen, G., Tewordt, L., 1959, *Phys. Rev.,* **113,** 982.
Benin, D., 1970, *Phys. Rev.,* **B1,** 2777.
Berman, R., Foster, E. L., Ziman, J. M., 1955, *Proc. Roy. Soc.,* **A231,** 130.
Bross, H., 1962, *Phys. Stat. Sol.,* **2,** 481.
Brown, R. A., 1967, *Phys. Rev.,* **156,** 692.
Callaway, J. C., 1959, *Phys. Rev.,* **113,** 1046.
Callaway, J. C., Boyd, K., 1964, *Phys. Rev.,* **134,** A1655.
Carruthers, P. A., 1961, *Rev. Mod. Phys.,* **33,** 92.
Casimir, H. B. G., 1938, *Physica,* **5,** 495.
Devyatkova, E. D., 1957, *Zhurn. Tekh. Fiz.,* **27,** 461.
Devyatkova, E. D., Moizhes, V. Ya., Smirnov, I. A., 1959, *Fiz. Tverdogo Tela,* **1,** 613.
Douglass, R. L., 1963, *Phys. Rev.,* **129,** 1132.
Drabble, J. R., Goldsmid, H. J., 1962, *Thermal Conduction in Semiconductors* (Pergamon Press, Oxford).
Dresden, M., 1961, *Rev. Mod. Phys.,* **33,** 265.
Elliott, R. J., Parkinson, J. B., 1967, *Proc. Phys. Soc.,* **92,** 1024.
Genzel, L., 1953, *Z. Physik,* **135,** 177.
Greenwood, D. A., 1962, *Proc. Phys. Soc.,* **80,** 226.
Griffin, A., Carruthers, P., 1963, *Phys. Rev.,* **131,** 1976.
Guyer, R. A., Krumhansl, J. A., 1966, *Phys. Rev.,* **148,** 766, 778.
Hamilton, R. A. H., 1973, *J. Phys. Chem.,* **6,** 2653.
Hamilton, R. A. H., Parrott, J. E., 1969, *Phys. Rev.,* **178,** 1284.
Harrison, W. A., 1970, *Solid State Theory* (McGraw-Hill, New York).
Herring, C., 1954, *Phys. Rev.,* **95,** 954.
Herring, C., 1967, *Phys. Rev. Letters,* **19,** 167.
Horie, C., Krumhansl, J. A., 1964, *Phys. Rev.,* **136,** A1397.
Ishioka, S., Suzuki, H., 1963, *J. Phys. Soc. Japan,* **18,** Suppl. II, 93.

Jones, H., 1956, in *Handbuch der Physik,* Ed. S. Flügge, Vol.19 (Springer Verlag, Berlin).
Joshi, A. W., Sinha, K. P., 1966, *Proc. Phys. Soc.,* **88**, 685.
Joshi, Y. P., Tiwari, M. D., Verma, G. S., 1970, *Phys. Rev.,* **B1**, 642.
Kadanoff, L. P., Martin, P. C., 1961, *Phys. Rev.,* **124**, 670.
Keyes, R. W., 1961, *Phys. Rev.,* **122**, 1171.
Kittel, C., 1970, *Introduction to Solid State Physics,* 4th edition (Wiley, New York).
Klein, M. V., 1969, *Phys. Rev.,* **186**, 839.
Klemens, P. G., 1955, *Proc. Phys. Soc.,* **A68**, 1113.
Klemens, P. G., 1958, *Solid State Phys.,* **7**, 1.
Korenblit, L. L., Sherstobitov, V. E., 1968, *Fiz. Tekh. Poluprov.,* **2**, 688.
Krumhansl, J. A., 1965, in *Proceedings of the International Conference on Lattice Dynamics, Copenhagen,* (Pergamon Press, Oxford) p.523.
Kwok, P. C., Martin, P. C., 1966, *Phys. Rev.,* **142**, 495.
Kwok, P. C., 1967, *Solid State Phys.,* **20**, 213.
Lynton, E. A., 1964, *Superconductivity* (Methuen, London).
McCollum, D. C., Wild, R. L., Callaway, J., 1964, *Phys. Rev.,* **136**, A426.
McIrvine, E. C., 1959, *Phys. Rev.,* **115**, 1537.
Madelung, O., 1957, in *Handbuch der Physik,* Ed. S. Flügge, Vol.20 (Springer Verlag, Berlin) p.58.
Maradudin, A. A., 1964, *J. Amer. Chem. Soc.,* **86**, 3405.
Maradudin, A. A., 1966, *Solid State Phys.,* **18**, 273.
Maradudin, A. A., Ipatova, I. P., Montroll, E., Weiss, G. H., 1971, *Theory of Lattice Dynamics in the Harmonic Approximation,* 2nd edition (Academic Press, New York).
Men', A. A., Sergeev, O. A., 1973, *High Temperatures - High Pressures,* **5**, 19.
Orbach, R., 1962, *Phys. Rev. Letters,* **8**, 393.
Parrott, J. E., 1971, *Phys. Stat. Sol. (b),* **48**, K159.
Pippard, A. B., 1955, *Phil. Mag.,* **46**, 1104.
Sato, H., 1955, *Prog. Theor. Phys. (Kyoto),* **13**, 119.
Schieve, W. C., Peterson, R. L., 1962, *Phys. Rev.,* **126**, 1456.
Simons, S., 1963, *Proc. Phys. Soc.,* **82**, 401.
Spector, H. N., 1966, *Solid State Phys.,* **19**, 291.
Srivastava, G. P., 1973, *Phys. Stat. Sol. (b),* **56**, K 39.
Steigmeier, E. F., Kudman, I., 1966, *Phys. Rev.,* **141**, 767.
Walker, C. T., Pohl, R. O., 1963, *Phys. Rev.,* **131**, 1433.
Wilson, A. H., 1953, *The Theory of Metals* (Cambridge University Press, Cambridge).
Ziman, J. M., 1954, *Proc. Roy. Soc.,* **A226**, 436.
Ziman, J. M., 1956, *Phil. Mag.,* **1**, 191.
Ziman, J. M., 1960, *Electrons and Phonons* (Oxford University Press, Oxford).

5

The analysis of experimental thermal conductivity data

5.1 Introduction

The theory of thermal conductivity, although not devoid of intrinsic intellectual interest, must be regarded primarily as a tool for the interpretation of experimental results, and, if possible, their prediction. In this chapter the material presented in chapter 4 will be turned to this end. Most of the discussion will concern analysis of data but prediction will not be forgotten and an attempt will be made to show the limits within which prediction can operate at present. However, the subject matter of the chapter will be limited to those materials which are relatively well behaved. Composites containing two or more phases and inhomogeneous materials in general will not be discussed here but in chapter 6.

The layout of this chapter is approximately in order of increasing complication. In nearly all cases the pure metals can be regarded as possessing a thermal conductivity entirely electronic in nature, and these are dealt with first. Again, at low temperatures insulators and almost all semiconductors show only lattice heat conduction, which can therefore be discussed in isolation in the following section. Metallic alloys show a situation where there is both lattice conduction and electronic conduction; however, the behaviour of the electrons is simplified by their degeneracy. This section introduces the problem of separating the two kinds of heat transport. Next the special case of superconducting metals is considered, followed by an account of the way in which thermomagnetic effects can be used to separate electronic and lattice conduction. The last interpretative section deals with the high-temperature behaviour of insulators and semiconductors with the attendant complications arising from the presence of not only electrons, holes, and phonons, but also of photons as carriers of heat. Finally the possibilities and limitations in the prediction of thermal conductivity are considered.

It is worth stressing in this introduction that the approach to free carrier thermal conductivity and that to phonon thermal conductivity are quite different. In the case of the electrons (and/or holes) the problem is normally regarded as one of determining the Lorenz number, it being assumed that the electrical conductivity is known. This is generally a reasonable assumption since it is easier to measure the electrical conductivity than the thermal conductivity. If, however, the electrical conductivity is not known, then one is essentially in the position of needing to predict it theoretically. This is a problem of a much higher order of difficulty than that of determining the Lorenz number. If the electrical conductivity is known, say, at room temperature, then it is relatively easy to predict its value at high temperatures, but at low temperatures the effects of an unknown residual resistance will become important. These remarks apply to metals; in semiconductors it is more difficult still to predict electrical conductivity, since the number of free electrons may vary from one sample of the same material to another, or from one temperature to another. Thus, except for the case of insulators, empirical information on the electrical conductivity is essential if the thermal conductivity is to be successfully analysed.

The lattice heat conduction, on the other hand, has to be treated as an entity in itself. This does not mean that nothing can be done to predict it, but it does make it much more an object of sometimes rather speculative analysis. This difference means that, if the thermomagnetic method is not possible or is inconvenient for some reason, one normally regards it as best to calculate the electronic thermal conductivity as accurately as possible, so that the lattice conductivity is obtained by subtracting the electronic from the total conductivity.

It will be clear that the existence of certain methods of analysing thermal conductivity data should have its influence on the design of experiments whose purpose is not only to measure but also to understand the thermal conductivity of a particular material. This should be the objective in applied as well as pure research. Assuming that the material and the temperature range are given, and leaving aside those questions of technique discussed in chapters 2 and 3, what considerations should the investigator bear in mind? To begin with, if the material is an electrical conductor, it is highly desirable that he should determine the electrical conductivity over the same temperature range as the thermal conductivity. In the case of metals and alloys this is really quite essential. If the material becomes superconducting it is very useful to be able to quench the superconductivity with an applied magnetic field. Where high mobility semiconductors or semimetals are concerned, it is worth considering the possibilities outlined in section 5.6 of separating the electronic and lattice thermal conductivity by using the thermomagnetic effects. Supposing that one is able to determine the lattice thermal conductivity accurately and that the temperature range is appropriate, it is of great assistance to have a number of specimens covering a range of point defect concentrations, since, as shown in section 5.3, this makes possible a more critical assessment of the three-phonon scattering. These are a few of the theoretical factors which show how to make the most of the results of thermal conductivity experiments. Other examples appear in the following sections of this chapter.

5.2 Heat conduction in pure metals

The thermal conductivity of pure metals is almost entirely electronic and therefore the analysis is largely based on the theories discussed in sections 4.7 and 4.8 at least in the case of monatomic and other relatively well behaved elements. The situation is rather different in the case of the transition metals and rare earths which merit a brief discussion on their own.

In all cases it is extremely important also to have measurements of the electrical conductivity so as to test the Wiedemann–Franz law. Furthermore it is important, particularly at low temperatures, that the thermal and electrical conductivity be determined on the same specimen, since the effects of impurities and lattice defects will vary from one sample of the same metal to another. This is not so necessary at high temperatures, where lattice scattering predominates.

Although it is usually assumed that lattice heat conduction is negligible in monatomic metals, there is some evidence that this is not always so. For instance Stauder and Mielczarek (1967) found what appears to be a large lattice component in potassium. Now, the almost spherical Fermi surface of potassium might be taken to mean that the simple theory by which only longitudinal phonons interact with electrons holds in this case, so that the transverse phonons might support a large thermal conductivity. Stauder and Mielczarek's results were obtained at low temperatures, but it is not unusual to find that the high-temperature Lorenz ratio $\lambda/\sigma T$ shows a small discrepancy from the standard value of $\pi^2 k_B^2/3e^2$ given in equation (4.48). There is, however, no universal and unambiguous interpretation of these small differences.

Further discussion of phonon heat conduction in metallic conductors will be given in section 5.4 devoted to alloys.

The normal pattern of the temperature dependence of the electronic thermal conductivity λ_e can be seen from figure 5.1. It is characterised by an independence of temperature at temperatures higher than about $\frac{1}{2}\Theta_D$; at lower temperatures λ_e begins to increase owing to the decreasing strength of the lattice scattering. Eventually λ_e passes through a maximum and begins to decrease in accordance with the behaviour to be expected when static imperfection scattering predominates.

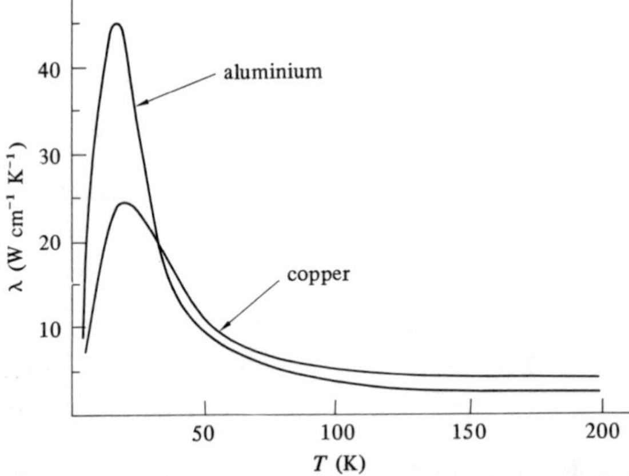

Figure 5.1. Normal pattern of the temperature dependence of the electronic thermal conductivity.

By referring to equations (4.53) and (4.63) it will be seen that the expected form of the electronic thermal resistivity at low temperatures will be

$$W = \alpha T^2 + \frac{\beta}{T}, \tag{5.1}$$

where $W_i = \alpha T^2$ is the intrinsic or ideal resistance and $W_0 = \beta/T$ is the residual resistance. Thus it is convenient to plot experimental data as WT against T^3 (Rosenberg, 1955), so that if the results do give a straight line relationship α and β may be readily ascertained. The value of β can be compared with $\rho_0/\mathcal{L}_0$ to which it is normally found to closely correspond. Failure to do so is usually regarded as evidence of a breakdown of Matthiessen's rule. Once β has been determined, W_i can be obtained even if it is not of the form αT^2, though with diminishing accuracy as the temperature decreases and W_i becomes a smaller and smaller part of the total resistivity.

The value of the power of the temperature in W_i frequently exceeds two even in quite simple metals like copper. A possible explanation of this can be seen in equation (4.58) as arising from the first or third terms in the expression for the intrinsic thermal resistance.

It is possible to relate α to the high temperature thermal resistance W_∞ by combining equations (4.62) and (4.63) with (5.1) to give

$$\alpha = \frac{12 \times 5!\,\zeta(5)}{\pi^2} \frac{W_\infty}{\Theta_D} \left(\frac{k_F}{q_D}\right)^2 . \tag{5.2}$$

In practice, on using measured values of W_∞, α is usually too small by a factor of about four. The explanation here lies in the fact that the derivation of equation (4.58) neglected umklapp processes, which will increase the high-temperature thermal resistance relative to its low-temperature value. If the quantity $\alpha\Theta_D^2/W_\infty$ is compared for metals in the same group of the periodic table, it is commonly found to have a similar value. This is what might be expected because the parameter k_F/q_D depends on the number of electrons per atom, that is on the valency.

The maximum in W predicted by equation (4.58) is not observed, and again this is due to the effects of the neglected umklapp processes. Thus it will be seen that, where discrepancies arise between high-temperature and low-temperature predictions, it is most likely the high-temperature result which is wrong.

White and Woods (1959) have found an empirical relation which is obeyed for a large number of metals, including transition metals:

$$\frac{W_i}{W_\infty} = 2\left(\frac{T}{\Theta_D}\right) \mathcal{J}_3\left(\frac{\Theta_D}{T}\right) , \tag{5.3}$$

but there does not appear to be any underlying significance in this result.

The transition metals show a number of differences from the behaviour described above. Firstly, at high temperatures the measured Lorenz ratio often considerably exceeds the Sommerfeld value. Secondly, the low temperature intrinsic Lorenz ratio does not appear to fall continuously as predicted by the simple theories of sections 4.7 and 4.8.

The behaviour at moderate to high temperatures has been interpreted in some cases as due to lattice conduction, as for instance in the work of Bäcklund (1961) on iron and subsequently on tungsten and molybdenum. Bäcklund adopts a special procedure to determine the electronic thermal conductivity:

$$\lambda_e = \frac{\mathcal{L}_0 T}{\rho + \bar{\rho}_0} , \tag{5.4}$$

where $\bar{\rho}_0$ is the absolute value of the intercept on the ρ axis obtained when linear ρ against T behaviour (after subtraction of residual resistance) is extrapolated to absolute zero. He then takes $\lambda_{ph} = \lambda - \lambda_e$. Considerable reservations must be felt about these results at other than rather high temperatures, however.

On the other hand Goff (1970) finds the high temperature Lorenz ratio of chromium is impossible to understand in terms of lattice conduction and views it as a consequence of the changes in electron density of states near the Fermi level. It is really not possible in the present state of the theory to provide a completely convincing explanation of the high-temperature thermal conductivity of the transition elements.

At low temperatures the situation appears to be more satisfactory. A series of investigators (e.g. Schriempf, 1968), suggested that, instead of the low temperature intrinsic Lorenz ratio, $\rho_i/W_i T$, decreasing continuously as the temperature decreases, it appeared to be tending to a constant of about one third the Sommerfeld value. This was explained as due to electron–electron collisions by Herring (1967), and, although there was a modest numerical discrepancy between theory and experiment, this is confirmed by the temperature dependence of both W_i and ρ_i.

One general characteristic of the transition metals is that their thermal conductivity is lower than that of ordinary metals. This is even more noticeable for the rare earth and actinide metals, which also share with the transition metals the presence of anomalies in W and ρ associated with various magnetic transitions, where these occur. However, generally there is no anomaly in the Lorenz number.

For the rare earths the Lorenz ratios are commonly reported as exceeding the Sommerfeld value, but the reported results vary as between single crystals and polycrystals, and there is some evidence that the nonmagnetic rare earths have normal Lorenz ratios. The picture is further confused by the fact that, in addition to lattice conduction, heat transport by magnons seems a distinct possibility, though in no case has this been definitely established as occurring. Results on a number of rare earths have been recently reported by Nellis and Levgold (1969).

5.3 Low-temperature lattice thermal conductivity of semiconductors and insulators

The absence of electronic effects brings a considerable simplification to the analysis of thermal conductivity data. Such is the case for insulating crystals and semiconductors having negligible concentrations of donors or acceptors. In this section we consider their low-temperature behaviour, where by low temperatures we mean temperatures less than about one half of the Debye temperature. The materials actually referred to will be largely the alkali halides, group IV and III–V semiconductors, together with the important case of solid helium.

With some of these materials it is reasonable to hope that their purity and crystalline perfection is so great that only boundary scattering and three-phonon processes will be effective in giving rise to thermal resistance. In particular, it has proved possible to prepare almost monoisotopic crystals of lithium fluoride, helium (^{3}He and ^{4}He), and germanium. In other cases it has to be admitted that residual impurities play a significant role at temperatures close to that of the thermal conductivity maximum. Yet again, there have been many experiments performed with samples of predetermined isotopic or chemical composition. Finally the effects of dislocations and precipitated colloidal particles have been investigated.

The starting point for the theoretical analysis of lattice thermal conductivity will normally be the relaxation time equations in section 4.6, especially the Debye and Callaway formulae. However, it is not uncommon to find that research workers have turned to more sophisticated forms of analysis in the attempt to explain their results more exactly. If we leave this latter possibility until later, the important question is whether the simpler Debye formulation is adequate or is it necessary to take advantage of the more subtle treatment of the N-processes provided by Callaway? The Guyer–Krumhansl formula has not received much application and indeed can be seen in some cases to give an erroneous result.

Broadly speaking one would expect that if the thermal conductivity is relatively low then the Debye equation will suffice, but if it is high then Callaway's method must be used. Harrington and Walker (1970) quote (without further reference) a rule of thumb that if the peak value of the thermal conductivity is less than 10 W cm^{-1} K^{-1} then the Debye equation will certainly be adequate but if it is greater than 100 W cm^{-1} K^{-1} it will certainly be necessary to use the Callaway formula.

The frequency and temperature dependence of the different N- and U-process relaxation times are summarised in table 4.1. Which form to actually use is often a difficult question to settle, particularly if, as is often the case, the Debye or Callaway equations are used without distinguishing between the different polarisation branches. The most commonly used form for N-processes is, in fact, not in the table and makes $\tau^{-1} \propto \omega^2 T^3$. In the case of U-processes most of the temperature variation comes from the exponential term, so the power of T does not matter much; once again the most commonly used frequency dependence is $\tau^{-1} \propto \omega^2$. In the event of failing to obtain a convincing fit using these forms for τ^{-1}, many authors have not hesitated to regard the powers of ω and T as adjustable parameters. It is sometimes possible to draw conclusions about the nature of the scattering in this way.

A slightly spurious simplification often arises because of the fact that the strength of both the N- and U-process scattering is used as an adjustable parameter. This means that the distinction between the Debye equation (4.32) in which N-processes are neglected and the first Callaway term in (4.37) which includes N-processes often disappears in the variation of these adjustable parameters. This is probably the reason why the workers concerned with isotopic mixtures find it necessary to use the full Callaway formula, since consistency requires them to use the same strength of N- and U-process scattering for all their samples, whilst the mass difference scattering strength varies very considerably. It is probably correct to say that the best way to proceed is to fit the data using Callaway's equation and to revert to the Debye form if Callaway's correction term has been shown to be small.

One place where the normal processes play a particularly curious role is in the very low-temperature boundary scattering regime. Provided they are not strong enough to cause Poisseuille flow they can be ignored. It is then to be expected that the simple T^3 relation will be found and indeed it is. Recently Hurst and Frankl (1969) have carried out a detailed study of boundary scattering in silicon. One unexpected result was the existence of some specularity in apparently quite rough surfaces and that this appeared to be a function of heat flow direction rather than surface orientation. For purely diffuse scattering they give reasons why in rectangular specimens of edge lengths L_1 and L_2 the mean free path should be an harmonic mean

$$\frac{1}{\ell} = \frac{1}{2}\left(\frac{1}{L_1} + \frac{1}{L_2}\right),$$

rather than the expression

$$\frac{1}{\ell} = \frac{1}{2}\left(\frac{\pi}{L_1 L_2}\right)^{1/2}$$

used hitherto. The new expression seems in slightly better agreement with experiment. A special form of average velocity must be used:

$$\bar{v} = \frac{\sum_s v_s^{-2}}{\sum_s v_s^{-3}}$$

if it is desired to use the $\lambda_{ph} = \frac{1}{3}Cv\ell$ expression.

The only occasion on which Poisseuille flow has been reported was in the work of Mezhov-Deglin (1965) on solid helium. The theory of this is covered by equation (4.42); the main characteristic is that the thermal conductivity increases more rapidly than T^3, possibly up to T^8, over a fairly narrow range of temperature. It requires an extraordinary degree of purity (in the monoisotopic as well as chemical sense) for the effect to be observed. The reason why this is possible in the case of helium is that the different physical properties of the two helium isotopes greatly assist their separation.

The great purity possible in helium specimens also enables the unambiguous observation of U-process scattering. Furthermore the Debye temperature can be varied by altering the pressure under which the helium is solidified. Thus Seward *et al.* (1969) were able to prepare samples of ^{4}He with Debye temperatures in the range $48 \cdot 5 < \Theta_D < 95$ K and plot all the data covering four decades on a common curve of $\log\lambda$ against Θ_D/T showing that

$$\lambda \propto \exp(\Theta_D/bT)$$

with $b \approx 2 \cdot 13$. The fact that observations of a similar kind are possible in ^{3}He, makes the results of measurements on isotopic mixtures particularly interesting. Agrawal (1967) included the following scattering mechanisms in a full Callaway analysis

(i) boundary scattering, $\tau_B^{-1} = v/\ell$;
(ii) isotope scattering, $\tau_D^{-1} = A\omega^4$;
(iii) N-process scattering, $\tau_N^{-1} = B_N T^3 \omega^2$;
(iv) U-process scattering, $\tau_U^{-1} = B_U T^3 \omega^2 e^{-a/T}$.

The result of his curve fitting was that (i) in the neighbourhood of the thermal conductivity maximum the second term in the Callaway formula was dominant, and (ii) the isotope scattering was many times stronger than predicted by equation (4.24). The reason for this last result lies in the fact that in solid helium the zero-point energy vibrations are very large producing what is, in effect, a difference in the interatomic binding forces between ^{3}He and ^{4}He.

A great deal of research has been carried out on the thermal conductivity of the alkali halides. Berman and Brock (1965) studied the properties of lithium fluoride containing varying amounts of ^{6}Li and ^{7}Li (fluorine possesses only one isotope). They used the theoretical equation for the isotope scattering but modified the relaxation time for normal processes to $\tau_N^{-1} = B_N T^3 \omega$ which gave a much better fit. As with the helium analysis, the correction term in the Callaway formula was predominant in the neighbourhood of the thermal conductivity maximum (figure 5.2). It was impossible to fit the data without this term.

Most of the work on alkali halides has concerned the effects of deliberately added chemical impurities or lattice defects. The Debye

formulation, generally without an N-process contribution, has formed the basis of the analysis. The results can be summed up as follows:
(i) Resonance scattering generally occurs in the presence of point defects, but the theoretical expressions for the relaxation time, exemplified by equation (4.55), have generally proved inadequate. An exception was the use of Krumhansl's formula by Schwartz and Walker (1967) whose paper in addition contains a useful discussion of the problems of curve fitting where resonance scattering occurs. Brown and Popovic (1972) describe a computer program for this purpose.
(ii) Where impurities have precipitated as colloidal particles, the equation given in section 4.5, $\tau_{\cdot}^{-1} = N\sigma v$, seems to satisfactorily describe the situation (Worlock, 1966).
(iii) Dislocations generally seem to produce a relaxation time whose frequency dependence agrees with theory [equations (4.27) and (4.28)], but whose size is orders of magnitude smaller (Moss, 1966). A similar but less drastic situation occurs in the lattice conductivity of metallic alloys (see section 5.4).

Presumably the success of the Debye formula under these conditions is due to the presence of strong additional scattering, rather than to any special feebleness of the N-processes in the substances studied.

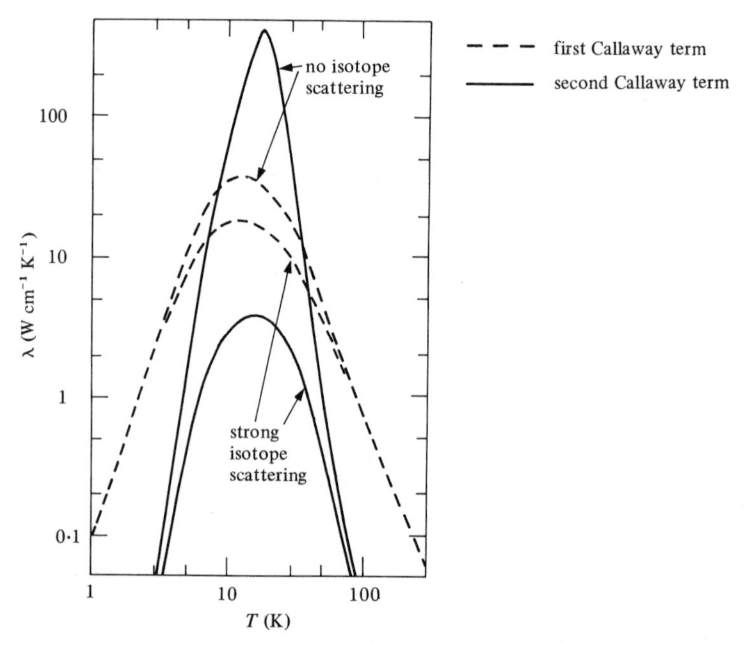

Figure 5.2. Theoretical curves used to fit data on lithium fluoride. [Berman and Brock, 1965, figure 6.]

The case considered in Callaway's original paper (1959) was that of germanium; his results for natural and isotopically enriched material are shown in figure 5.3. There have been a certain number of similar analyses since, but to a great extent the analysis of semiconductor data has tended to take a different turn, due to the work of Holland (1963) and Bhandari and Verma (1965). These authors have considered it important to treat the contribution of the transverse and longitudinal modes separately, whilst neglecting the Callaway correction term. In addition they divide the transverse branch into two parts having different group velocities, so as to allow for the very strong dispersion of this branch in germanium. This results in an absence of umklapp scattering for low-frequency transverse phonons. If this method is used, an excellent fit is obtained to the experimental results; this is shown for germanium in figure 5.3. Also shown are the results of the variational calculation of Hamilton and Parrott (1969). Both this and Holland's calculation agree in finding that transverse phonons generally play a preponderant role in heat conduction.

There have been a number of other models employed in thermal conductivity calculations, generally with a view to taking proper account of the dispersion of the lattice vibrational frequencies which affects both the density of states $g(\omega)$ and the group velocity of the phonons. However, there is a feature which these analyses share with those in the previous paragraph and which must be regarded as unsatisfactory. This is a multiplication of the number of adjustable parameters, which leaves the reader with a feeling that the fact of agreement with experiment may not be very significant as regards either the underlying theory or the numerical values of these parameters.

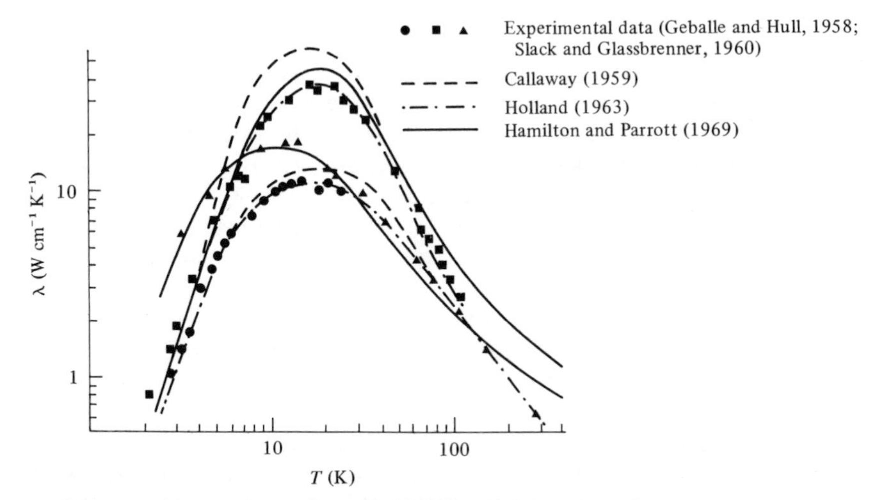

Figure 5.3. Comparison of experimental data on germanium with theoretical calculations.

For more heavily doped semiconductors it will be necessary to allow for electronic (or hole) heat conduction, as well as scattering of phonons by electrons, holes, and neutral donor acceptors. Actual heat conduction by free carriers is nearly always relatively small and can be separated from the lattice heat conduction with the use of the methods of section 5.7. Again the scattering of phonons by electrons can be adequately treated by means of equation (4.29). Scattering by neutral donors and acceptors has, however, caused great difficulties in the past, and indeed even now there is no proper theory of the effects of neutral acceptors.

One class which requires special comment is that of highly anisotropic materials, of which graphite is the most important member. Crystallographically we have a layer structure with tight interatomic binding within the layers and very weak interlayer binding forces. It may be found that there are effectively two Debye temperatures for vibrations within and normal to the layer structure; in the case of graphite the in-plane value is ~2500 K and the out-of-plane value 180 K. Similarly, the thermal conductivity will be very anisotropic. The conductivity in the layer λ_a may be more than an order of magnitude greater than the component λ_c normal to the layer structure. This means incidentally that, owing to slight misalignment, an ostensible measurement of λ_c often contains a considerable contribution from λ_a.

A straightforward approach to the analysis of the thermal conductivity of graphite has been described by Taylor (1966). It is worth noting that, although his measurements went up to 900 K, this is a 'low temperature' for λ_a, though not for λ_c. A two-dimensional model is used for λ_a in the form

$$\lambda_a = \tfrac{1}{2} C_v v_s \ell$$

where the heat capacity and sound velocity are defined appropriately for such a two dimensional structure. The mean free path depends on boundary and umklapp scattering, the latter being identified by plotting the measured ℓ logarithmically against $1/T$. The experimental data for λ_a were very well accounted for in this way. λ_c on the other hand was compared by Taylor with the Leibfried–Schlömann formula to be discussed in section 5.7; semiquantitative agreement was obtained. For a recent review of this subject see Kelly (1973).

To conclude this section, a brief description will be given of the treatment of amorphous materials such as glasses and noncrystalline polymers. In section 4.5 a simpler theoretical approach was shown to predict that the atomic disorder would give a reciprocal relaxation time proportional to q^2 (or ω^2) at low frequencies but independent of frequency at high frequencies, with a smooth transition between these two regimes. Some typical experimental results for quartz glass (Berman, 1951) and polymethyl methacrylate (Choy *et al.*, 1970), are shown in figure 5.4. The very small value of the thermal conductivity is of course to be expected. The temperature dependence has been accounted for by Dreyfus *et al.* (1968), who have somewhat extended the simple theoretical picture alluded to above. There is in amorphous materials an excess specific heat which is attributed to a band of localised modes of vibration. This gives rise to a resonant scattering of phonons which has to be added to the disorder scattering. When this is done a good fit can be made to the experimental data (see figure 5.4 again) except for some very low temperature measurements of Choy *et al.* (1970).

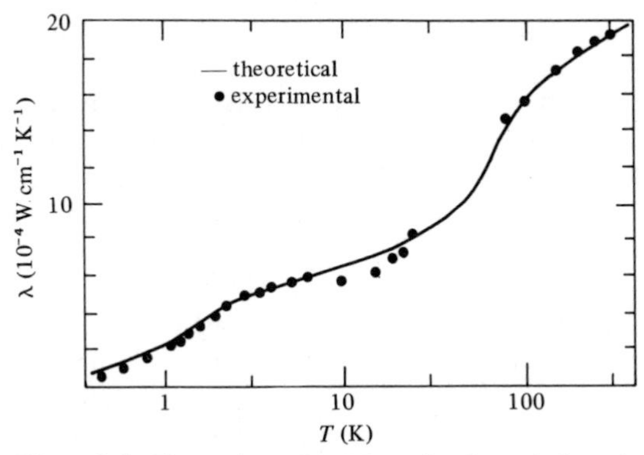

Figure 5.4. Thermal conductivity of polymethyl methacrylate. [Choy *et al.*, 1970, figure 2.]

5.4 The thermal conductivity of metallic alloys

There have been a large number of investigations of the thermal conductivity of metallic alloys, almost exclusively below room temperature. The scientific interest arises from the possibility of determining the lattice thermal conductivity in a metallic conductor to a degree of accuracy not possible in the case of pure metals. A number of interesting problems which have arisen in the course of this work will be touched on in this section.

The basic question concerns the separation of the thermal conductivity into electronic and lattice (phonon) contributions,

$$\lambda = \lambda_e + \lambda_{ph}.$$

The factors involved in calculating λ_e with the assistance of data on the electrical conductivity have been discussed in the section on heat conduction in pure metals. At high temperatures ($T > \frac{1}{2}\Theta_D$), for those alloys where the solvent metal obeys the Wiedemann–Franz law, it will be safe to assume the same for the alloy. However, as the temperature increases, the accuracy with which λ_{ph} can be estimated decreases, because, while λ_e is independent of temperature, λ_{ph} is proportional to T^{-1}.

At low temperatures there is the difficulty that the Wiedemann–Franz law breaks down badly for the intrinsic resistance. The situation is less difficult, however, for alloys than for pure metals, for two reasons.

Firstly the static imperfection scattering is greater, so errors in estimating W_i are less significant, and secondly, for dilute solid solutions, it will be possible to use the value of W_i obtained for the pure solvent metal. For these reasons it is usually possible to determine λ_{ph} quite accurately at liquid helium temperatures.

The electronic thermal conductivity itself is not of any great interest in the study of alloys. One point that has been investigated concerns the applicability of Matthiessen's rule that resistivities are additive. It appears from the work of Farrell and Grieg (1969) that the deviations from Matthiessen's rule in nickel alloys are larger for thermal resistance than for electrical resistance.

A large part of the interest in the lattice thermal conductivity of alloys arises from the fact that it is possible to study electron scattering of phonons in these materials. The simple theory enshrined in equations (4.29) and (4.30) allows only for the scattering by electrons of longitudinal phonons. The assumptions involved in this theory almost certainly break down to the extent that some scattering of transverse phonons will occur, though this might be rather small in simple materials like the alkali or noble metals. If there were in fact negligible direct scattering of transverse phonons, then one would have to rely, at low temperatures, on normal processes to transmit the effects of scattering of longitudinal phonons to the transverse. Thus in Callaway's equation (4.37) the longitudinal phonons would appear in the first term and the transverse in the second. In practice it is frequently assumed that the transverse phonons are scattered by electrons just as strongly as the longitudinal.

If equation (4.30) is used in the thermal conductivity formula (4.32), it will be seen that the ω^{-1} dependence of τ means that $\lambda_{ph} \propto T^2$. It can be shown that

$$\lambda_{ph} = \frac{\mathscr{C} T^2}{W_\infty \Theta_D^2 N(\mathscr{E}_F)} , \tag{5.5}$$

where W_∞ is the electronic thermal resistivity at high temperatures (where it is independent of temperature) and $\mathscr{C}$ is a constant which depends amongst other things on the assumptions made about the phonon–electron coupling discussed in the last paragraph. It is usually impossible to determine λ_{ph} for a pure metal but it can be determined for quite dilute alloys and extrapolated to the pure metal. The values of λ_{ph} obtained with different solutes agree quite well, but λ_{ph} varies surprisingly rapidly for very small concentrations of solute. It appears that this arises from changes in $N(\mathscr{E}_F)$, i.e. changes in band structure (Sousa, 1968).

A major complication is caused by the effects of electron mean free path, as first pointed out by Pippard (1955). These effects occur when this mean free path is smaller than the phonon wavelength, and in the limiting case should result in $\lambda_{ph} \propto T$ rather than T^2. This will tend to occur where the residual resistance is very high; indeed Lindenfeld and Pennebaker (1962) were able to show that for a given solvent metal $\lambda_{ph}/\rho_0 T$ would be a universal function of T/ρ_0, independent of the solute atom. These remarkable results are shown in figure 5.5. However the proportionality of λ_{ph} to T has not been observed, as it is masked by the effects of boundary scattering. As noted by Lindenfeld *et al.* (1966) this is a case where the simple addition of thermal resistivities gives very erroneous results which markedly underestimate the effects of boundary scattering.

Another effect which gives a thermal conductivity proportional to T^2 is scattering by dislocations. There have been a number of papers on this subject, for example by Lomer and Rosenberg (1959) and Charsley *et al.* (1968). The dislocations are put into the alloys by plastic deformation and their effect can be determined by subtracting WT^2 in the undeformed material from the same quantity in the deformed alloy to give $W_d T^2$ where W_d is the thermal resistivity due to dislocations. One problem here is the determination of the dislocation density N_d. The method employed by Charsley *et al.* was to cut up the sample into slices which were thinned down for electron microscopic observations of the dislocations. There are a number of methods of obtaining the dislocation density from the electron micrographs; the results of these determinations were found to agree to within 10%. From these dislocation densities a measure of the scattering per dislocation $W_d T^2/N_d$ can be obtained. In the copper–aluminium alloys examined the scattering power increased with aluminium

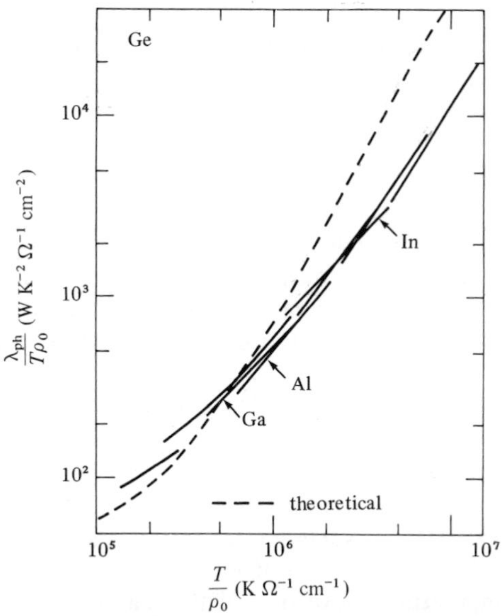

Figure 5.5. Universal curve for the lattice thermal conductivity of copper with germanium, gallium, aluminium, and indium as impurities. [Lindenfeld and Pennebaker, 1962, figure 7.]

content but in α-brass it was independent of zinc concentration. For vanishingly small aluminium concentration the value of $W_d T^2/N_d$ was around 6×10^{-8} cm^3 K^3 W^{-1}. The theoretical value can be obtained from equations (4.27) and (4.28) together with (4.32) and is about 4×10^{-9} cm^3 K^3 W^{-1}, somewhat more elaborate calculations can raise the figure to about 7×10^{-9} cm^3 K^3 W^{-1}. It will be seen that the experimental figure is an order of magnitude higher. Even larger discrepancies are found for dislocation scattering in insulators (see section 5.3). At present there is no convincing explanation for this.

Alloys must contain a considerable amount of lattice disorder so it is not surprising that scattering by point defects can become manifest at higher temperatures. Examples of this have been reported by Archibald *et al.* (1967) for potassium–caesium alloys, van Baarle *et al.* (1967) for silver based alloys and Srivastava *et al.* (1970) for copper and silver alloys. However, none of the analysis carried out is as precise as that used for nonconducting solid solutions.

Finally, at still higher temperatures three-phonon scattering becomes the most prominent resistive process. Although the accuracy of the lattice thermal conductivity determinations in this temperature range is not very high, White (1960) was able to quote values of λ_{ph} for copper, silver, and gold which were in quite good agreement with the Leibfried–Schlömann equation (5.11).

5.5 Heat conduction in the superconducting state

The kind of changes in heat transport which occur when a metal becomes superconducting have already been suggested in sections 4.5 and 4.8. Below the superconducting transition temperature T_c some of the electrons condense into a zero-entropy state separated from the normal state by an energy gap. The energy gap increases as the temperature is decreased below T_c, and consequently an increasing fraction of the electrons are to be found in this special state where they can neither transport heat nor scatter phonons. The application of a magnetic field of a suitable magnitude will restore the metal to normal conduction. A general account of all these effects, together with a description of the important distinction between superconductors of the first and second kind (type I and type II) will be found, for example, in Lynton's (1964) monograph. A more extensive account of the thermal conductivity of superconductors is given by Mendelssohn and Rosenberg (1961).

The general character of the temperature dependence of thermal conductivity in the normal, λ_n, and superconducting, λ_{sc}, states is shown for lead and tin in figure 5.6. In the case of lead the superconducting transition occurs at a temperature above that of the maximum in λ_n so that the major 'normal' scattering mechanism is by phonons. For tin, on the other hand, because T_c is below the temperature of the maximum in λ_n, the predominant scattering is by static crystalline imperfections.

Due to the increasing condensation of the electrons below T_c we expect the λ_{sc} to decrease rapidly to the extent to which the normal heat conduction is due to electrons. That part of the heat conduction due to phonons, on the other hand, will tend to increase as the scattering by electrons is eliminated. Because of this it is perfectly possible for λ_{sc} to be greater than λ_n.

The first step in the analysis of the thermal conductivity of a superconductor is to determine the situation in the normal state by means of measurements in a sufficiently strong magnetic field. Can it be established whether there is any significant lattice conduction? In the case of the electronic conduction, is the scattering entirely of the residual kind or must phonon scattering be allowed for? These problems are discussed in the sections on metals and metallic alloys (sections 5.2 and 5.4). The most straightforward case is where at temperatures below T_c the normal conductivity obeys the Wiedemann–Franz–Lorenz law so that we have no lattice conduction and no phonon scattering of the electrons. We then expect that the Bardeen–Rickaysen–Tewordt (BRT) formula (4.64) will apply; in figure 5.7 we show Satterthwaite's (1962) data on aluminium together with the theoretical curve. It will be seen that the agreement is excellent and there is no sign of any lattice conduction.

The considerable amount of verification of the BRT formula has meant that it is used to obtain the lattice heat conduction by subtracting the theoretical values of the electronic thermal conductivity in the superconducting state λ_{esc} from the experimental values of the total thermal conductivity λ_{sc}. In the case of alloys it will often be also possible to determine the lattice thermal conductivity in the normal state. This will be described at low temperatures by the Lindenfeld–Pennebaker (1962) equation and it will then be possible to test the calculations of Bardeen *et al.* of the reduction of electron scattering of phonons given in equation

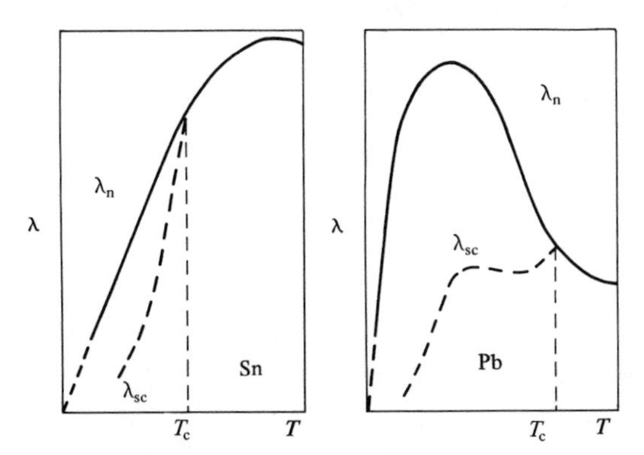

Figure 5.6. Thermal conductivity of tin and lead in the normal and superconducting states.

(4.31). The work of Sousa (1969) on tantalum–niobium alloys, figure 5.8, shows that excellent agreement between theory and experiment can exist.

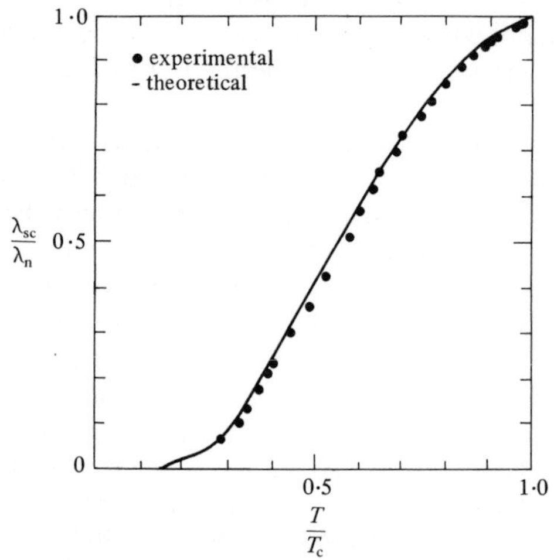

Figure 5.7. Thermal conductivity ratio λ_{sc}/λ_n for aluminium. [Satterthwaite, 1962, figure 3.]

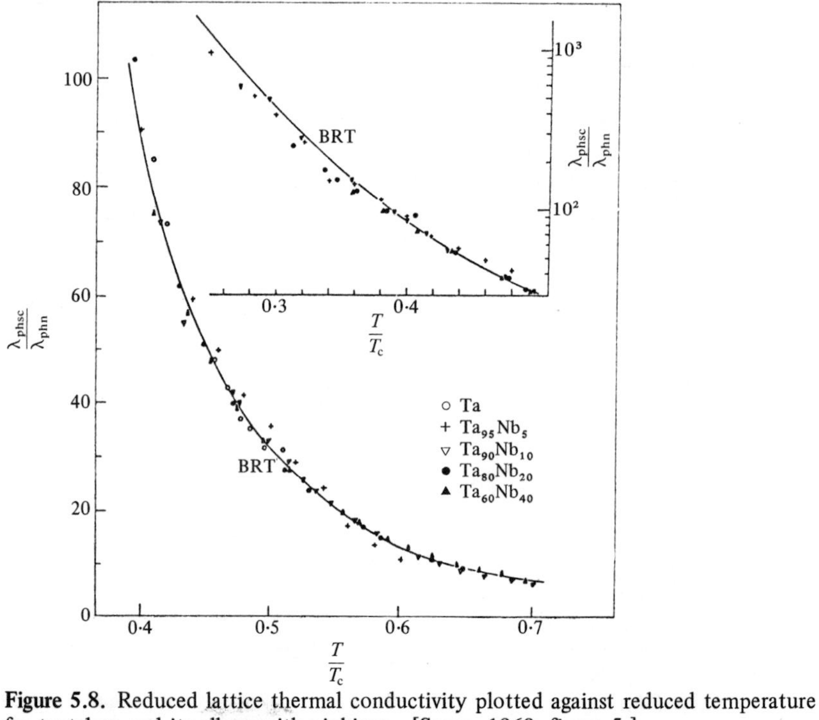

Figure 5.8. Reduced lattice thermal conductivity plotted against reduced temperature for tantalum and its alloys with niobium. [Sousa, 1969, figure 5.]

Due to the reduction of electron scattering, at sufficiently low temperatures it will be necessary to include other scattering mechanisms in the analysis of the lattice thermal conductivity. This should not present any special difficulties beyond those encountered normally in such analyses.

The situation is different for the case of λ_{esc} where an appreciable amount of the scattering of the normal electrons is by phonons. The theory for combined phonon and static-imperfection scattering has been worked out by Kadanoff and Martin (1961), but, because of the greater complication of the physical conditions, it has not proved possible to use it with the confidence felt for the BRT equations in the residual resistance case.

As well as in the rather straightforward situation for which the experimental results quoted above were obtained, measurements have been made in the intermediate or mixed state. In this state the conductor is divided into superconducting and normal lamellae; the boundaries between these give rise to scattering and consequently extra thermal resistance. Typical experimental work is that of Mendelssohn and Shiffman (1959); a recent attempt at theoretical interpretation has been made by Rickaysen (1968).

5.6 Thermomagnetic separation of electronic and lattice thermal conductivity

The problem of separating electronic and lattice thermal conductivity was discussed earlier in sections 5.4 and 5.5. In the case of alloys the Wiedemann–Franz law was invoked; this gives a satisfactory estimate of λ_e for the low temperature conditions where elastic scattering dominates, but difficulties arise where inelastic scattering becomes significant. Another method of analysis uses the thermal magnetoresistance, sometimes called the Maggi–Righi–Leduc effect. Applying a magnetic field enhances both the thermal and electrical resistivity and in some cases a large enough field will render λ_e quite negligible. Unfortunately in most cases these fields are unrealistically high. However, in certain situations this has proved a most successful method, particularly in the case of semiconductors and semimetals whose electrons have a high mobility.

The simplest way of considering the matter is to write down the total thermal conductivity in the form

$$\lambda(B) = \frac{\mathcal{L}T}{\rho(B)} + \lambda_{ph} \tag{5.6}$$

where $\rho(B)$ is the electrical resistivity in a magnetic field B. If then measurements of thermal conductivity and electrical resistivity are made for a number of values of magnetic field and the Lorenz number $\mathcal{L}$ is independent of B, a plot of λ against T/ρ extrapolated to $T/\rho = 0$ will give the lattice thermal conductivity λ_{ph}. It will be clear that for this method to succeed it will be necessary for fairly large changes of ρ to be produced by the magnetic field, at least by a factor of two. This method was applied to antimony by White and Woods (1958) and the results at two temperatures are plotted in figure 5.9. In view of the linear relationship between λ and T/ρ it seems likely that in this case, $\mathcal{L}$ was indeed independent of magnetic field.

A more sophisticated approach has been recently employed by Armitage and Goldsmid (1969) in an investigation of cadmium arsenide, Cd_3As_2. Cadmium arsenide is a semiconductor which has a high electron mobility and is always found as a degenerate n-type conductor. Armitage and Goldsmid used, in addition to the thermal magnetoresistance, the Righi–Leduc effect in their analysis. This effect consists of the production of a temperature gradient at right angles to an imposed temperature gradient by applying a magnetic field orthogonal to both. They employed the theory of Korenblit and Sherstobitov (1968) briefly alluded to in section 4.8. The electron mobility μ_H can be obtained from measurements of the electrical conductivity and the Hall coefficient. Armitage and Goldsmid used a thermal mobility $\mu_T = \mathcal{L}\mu_H/\mathcal{L}_0$ to express the change in thermal conductivity $\Delta\lambda$ in the form

$$-\frac{B^2}{\Delta\lambda} = \frac{1}{\mu_T^2\lambda_e} + \frac{B^2}{\lambda_e} \ . \tag{5.7}$$

Thus a plot of $-B^2/\Delta\lambda$ against B^2 should be linear with a slope of $1/\lambda_e$ and an intercept $1/\mu_T^2\lambda_e$ (λ_e is here the electronic thermal conductivity in zero field). The Righi–Leduc coefficient Λ can be used in the form

$$-\frac{1}{\Lambda} = \frac{1+\lambda_{ph}/\lambda_e}{\mu_T} + \frac{R_H\lambda_{ph}B^2}{\mathcal{L}_0 T} \tag{5.8}$$

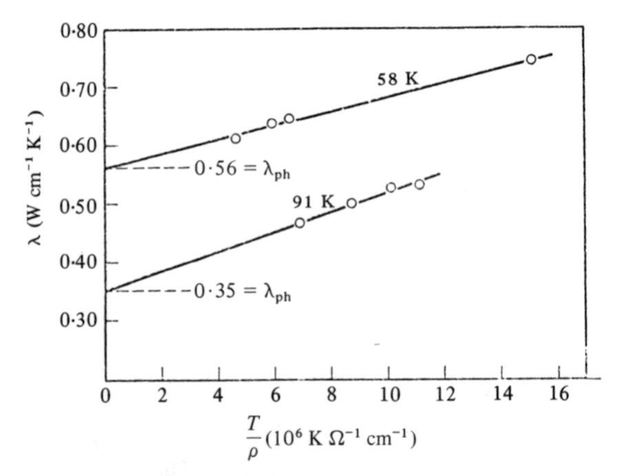

Figure 5.9. Thermal conductivity of antimony in magnetic fields of 0–14 kOe. [White and Woods, 1958, figure 3.]

where R_H is the Hall coefficient. Here the slope of a plot of $-1/\Lambda$ against B^2 will give $R_H\lambda_e/\mathcal{L}_0T$ and the intercept $(1+\lambda_{ph}/\lambda_e)/\mu_T$. Thus the measurement of the Righi–Leduc coefficient gives a valuable check on the results obtained from the thermal magnetoresistance. If the Hall coefficient is measured also, then the ratio $\mathcal{L}/\mathcal{L}_0$ can be obtained as well as λ_e and λ_{ph}. The results obtained by Armitage and Goldsmid are shown in figure 5.10; the values obtained for λ_{ph} and $\mathcal{L}/\mathcal{L}_0$ by the two methods were in good agreement. Since in some cases the Lorenz number was only about half the Sommerfeld value, an analysis assuming elastic scattering would have run into considerable difficulties; this underlines the value of the work of Korenblit and Sherstobitov. Furthermore, this treatment of the data does not require such large magnetoresistive effects as does the method used by White and Woods, described earlier.

Both methods assume that over the range of measurement the Lorenz number is independent of magnetic field, so that the second analysis has the advantage of needing relatively small fields for which this hypothesis is more plausible. Also, there is the general assumption that the lattice thermal conductivity is independent of magnetic field. This is of course known not to be the case where the material is either paramagnetic or

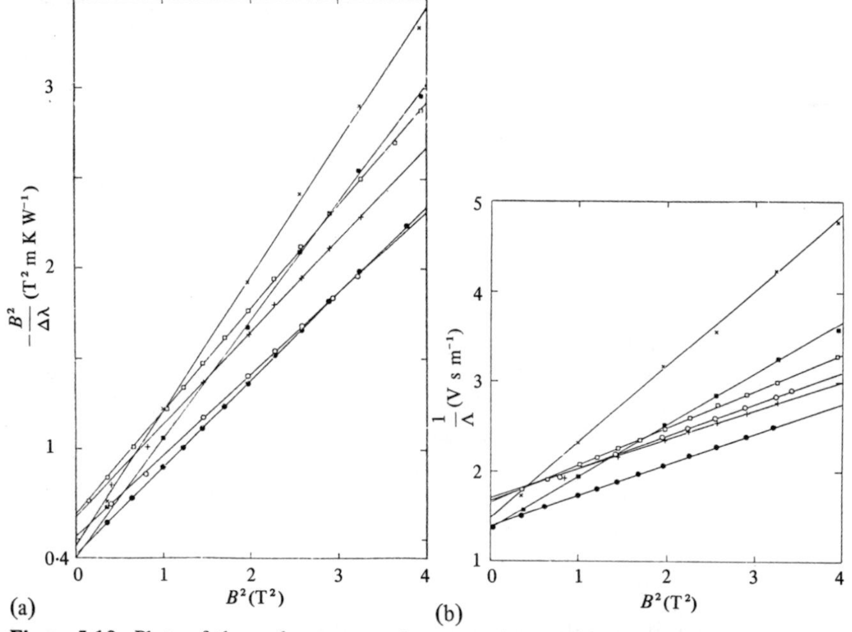

Figure 5.10. Plots of thermal magnetoresistance and the Righi–Leduc coefficient for different samples of cadmium arsenide at 300 K. [Armitage and Goldsmid, 1969, figures 1 and 2.]

contains a high concentration of paramagnetic impurity. It is possibly not the case where scattering of phonons by electrons is the major resistive mechanism, though there is no experimental evidence of this at present, nor is it expected on theoretical grounds. The assumption is therefore likely to be true for metals, alloys, and semiconductors. Finally, the methods described are not applicable to the case of mixed conduction described in equation (4.70). There the behaviour is rather complex and, in particular, the carrier thermal conductivity does not tend to vanish in the high field limit.

5.7 Insulators and semiconductors at high temperatures

The analysis of the thermal conductivity of insulators and semiconductors at high temperatures ($T > \frac{1}{2}\Theta_D$) may be a very complicated matter. The complexity arises in the first place from the possibility of simultaneous heat conduction by phonons, electrons and holes, and photons. Secondly, the theory of lattice thermal conductivity is complicated by the number of resistive mechanisms which may be at work. All this leads in many cases to an unavoidable lack of precision in the analysis.

It should also be remarked that the accurate measurement of thermal conductivity becomes more and more difficult as the temperature increases and it is not wise for a theoretician to adopt a too uncritical attitude towards the experimental data. Where a number of experimenters have measured the thermal conductivity of the same material large discrepancies often appear which depend systematically on the temperature.

Where sufficiently accurate data are available the procedure normally adopted is to consider first of all the photon contribution. It may often be possible to calculate the radiative heat conduction and subtract it from the total. To find the electronic thermal conductivity it is not likely to be possible to use the thermomagnetic methods outlined in section 5.6 because of low mobility of the electrons or holes. If, however, certain other electronic transport coefficients are measured, theoretical calculation of the electronic thermal conductivity is likely to be quite accurate. The lattice thermal conductivity then remains and can be subjected to such theoretical analysis as seems worthwhile. Incidentally, it will be rather unlikely that both radiation (photons) and electrons will make a significant contribution over the same temperature range.

The theoretical expression for the radiative thermal conductivity is given by equations (4.72) and (4.73); we reproduce (4.73) here

$$\lambda_r = \tfrac{16}{3}\sigma_R n^2 T^3 \langle \alpha^{-1} \rangle \,.$$

The temperature variation of λ_r arises in two ways, from the explicitly exhibited T^3 and through the temperature variation of $\langle \alpha^{-1} \rangle$. The most important contribution to $\langle \alpha^{-1} \rangle$ comes from the value of α in the neighbourhood of the peak of the blackbody radiation distribution. In an extrinsic semiconductor this will normally vary slowly with temperature causing λ_r to increase with T from a low value at around room temperature. However, at a certain temperature intrinsic conduction begins, the carrier concentration increases according to equation (4.50) and $\langle\alpha^{-1}\rangle$ begins to decrease very rapidly. This means that λ_r will exhibit a maximum at a particular temperature. If the low temperature electron or hole concentration is rather large, α may be also so large that the radiative thermal conductivity is always negligible. This is indeed usually the case.

As an example of radiative heat transfer, let us consider the results of measurements on two germanium–silicon alloys reported by Beers *et al.* (1962) and reproduced in figure 5.11. The rather less pure sample shows a smooth variation of λ with temperature, whilst the purer exhibits a peak. On the assumption that the difference was entirely due to λ_r, this quantity was obtained by subtracting one curve from the other. Since values of the absorption coefficient as a function of wavelength were available it was possible to calculate $\langle\alpha^{-1}\rangle$. The resulting theoretical points for λ_r are also shown in figure 5.11 and may be seen to be in good agreement with estimated experimental values. It is worth noting that no evidence of radiative heat conduction has been found in either pure silicon or germanium. In the case of germanium the smaller energy gap means that at temperatures high enough for the T^3 term to make enough impact $\langle\alpha^{-1}\rangle$ is very small. In silicon, on the other hand, where the energy gap is rather larger than in the alloy considered by Beers *et al.*, the thermal

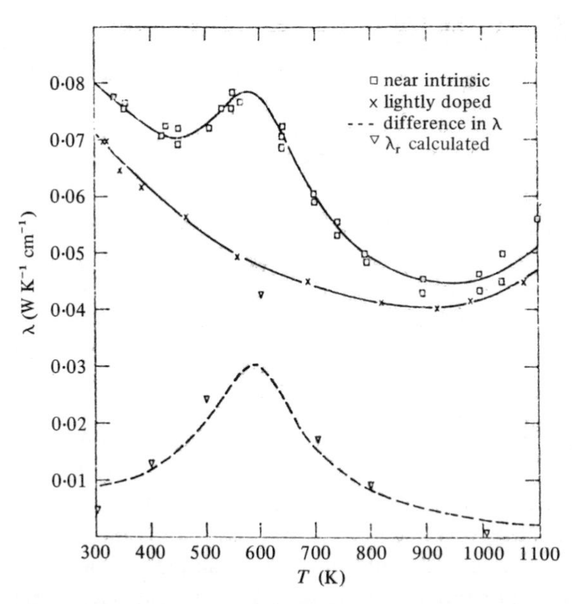

Figure 5.11. Effects of radiative heat transfer on the thermal conductivity of n-type germanium–silicon alloys. [Beers *et al.*, 1962, figure 4.]

conductivity is about an order of magnitude greater than in the alloy, so that the effect of radiative conduction might be difficult to observe. This view is supported by the fact that photon conduction has been clearly found in gallium arsenide, whose energy gap is larger than that of silicon, but whose thermal conductivity is distinctly lower. To sum up, a radiative contribution is most likely in high-energy-gap semiconductors of low lattice thermal conductivity; where its presence is suspected and optical absorption data are available, it will probably be found that it can be accounted for with the aid of theoretically calculated values. Incidentally, far larger intrinsic carrier concentrations are needed to give an appreciable carrier contribution to the thermal conductivity than are needed to eliminate the radiative heat flow, which is the reason why it is unlikely that both will be significant at the same temperature. In the case of insulators the optical absorption due to free carriers is absent, and therefore λ_r is able to increase continuously with temperature.

The effects of free electrons and holes have been observed in many semiconductors. The relevant theory is briefly discussed in section 4.8. What we shall discuss here is how this can be applied to experimental results, assuming for the present that relative confidence in the theories of free carrier and lattice thermal conductivity is such that it is sensible to calculate the former so as to leave the latter as a remainder. Provided some additional information on the electrical transport properties is available, this will normally be the best way to proceed.

Let us consider first the cases where only one type of carrier is present in any appreciable numbers. For an extrinsic semiconductor whose density of states $N(\mathcal{E})$ is given by equation (4.17) the equation for the Lorenz number (4.47) becomes

$$\mathcal{L}_0 \sim \left(\frac{k_B}{e}\right)^2 \left\{ \frac{(z+\frac{7}{2})F_{z+\frac{5}{2}}(\eta^*)}{(z+\frac{3}{2})F_{z+\frac{1}{2}}(\eta^*)} - \left[\frac{(z+\frac{5}{2})F_{z+\frac{3}{2}}(\eta^*)}{(z+\frac{3}{2})F_{z+\frac{1}{2}}(\eta^*)}\right]^2 \right\} , \tag{5.9}$$

where z is the power of the energy in the expression for the relaxation time (4.66) and $\eta^* = \mathcal{E}_F/k_B T$.

This of course assumes that the electron relaxation time is in fact describable in this way. Making this assumption we need the electrical conductivity σ, together with values of z and η^* to be able to calculate the electronic thermal conductivity. If η^* is large and negative, then $\mathcal{L}$ no longer depends on η^*, and the value of $\mathcal{L}$ is given by equation (4.68) and only z is needed additionally. Unfortunately in such cases λ_e is usually quite negligible so that in practice both η^* and z are likely to be needed. To determine these, measurements of the Seebeck and Hall coefficients are required since these coefficients also depend on η^* and z; if available, the value of the Nernst coefficient will provide added confirmation. A summary of the expressions for the Seebeck, Hall, and Nernst coefficients required for such a calculation is given by Putley (1960).

The procedure outlined in the previous paragraph is open to serious criticism on the ground that the total relaxation time is not likely to have the form given by equation (4.66), i.e. a simple power law of energy; this is particularly true for the more heavily doped semiconductors where impurity scattering of electrons will be present as well as the inherent scattering by phonons. Thus in their work on gallium arsenide, Amith *et al.* (1964) assumed the presence of both impurity and optical phonon scattering and were able to obtain the relative strength of the two scattering mechanisms, together with the reduced Fermi level, from measurements of Hall and Seebeck coefficients. They used the results to calculate the electronic thermal conductivity. Again, Steigmeier and Abeles (1964) were able to employ a similar technique in their study of heavily doped germanium–silicon alloys.

The situation in extrinsic semiconductors is considerably eased by the way in which our knowledge of the Lorenz number is likely to be most precise in those cases where the electronic thermal conductivity is largest. For conductors having relatively few electrons (or holes), for which therefore λ_e is small, the maximum range of $\mathcal{L}$ is $2(k_B/e)^2 < \mathcal{L} < 4(k_B/e)^2$, the actual value depending on the scattering mechanism, but as the number of carriers increases, and with it λ_e, $\mathcal{L} \rightarrow \frac{1}{3}\pi^2(k_B/e)^2$, no matter what type of scattering is present. These remarks do not apply, however, to intrinsic semiconductors; the limiting cases of metals and metallic alloys were discussed in section 5.2.

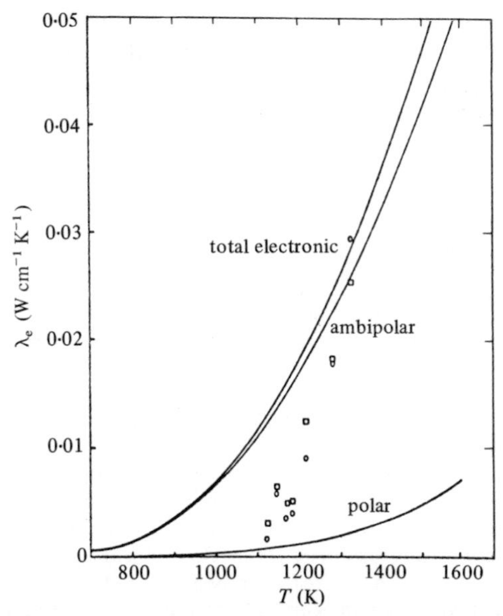

Figure 5.12. Calculated free carrier thermal conductivity of silicon. [Fulkerson *et al.*, 1968, figure 12.]

The position is indeed quite reversed in intrinsic semiconductors. Here the basic equation is (4.70) which, for fully intrinsic conduction (equal numbers of electrons and holes) and lattice scattering, becomes

$$\mathcal{L} = \left(\frac{k_B}{e}\right)^2\left[2+\frac{b}{(1+b)^2}\left(4+\frac{\mathcal{E}_g}{k_BT}\right)^2\right], \tag{5.10}$$

where b is the ratio of the hole mobility to the electron mobility. If this ratio is very different from unity, the second term becomes quite small. This equation was used by Fulkerson *et al.* (1968) in their study of the thermal conductivity of silicon up to 100°C. Their calculated free carrier thermal conductivity is shown in figure 5.12 and this was subtracted from the measured conductivity to obtain the lattice contribution.

Discussion of lattice thermal conductivity will be divided into three sections. Firstly we shall consider what might be called the intrinsic thermal resistance, which is always present and is due, in general terms, to anharmonic coupling between phonons; secondly the effects of atomic disorder, in particular the situation in solid solutions; and thirdly the contribution of the scattering of phonons by free electrons to the thermal resistance. These are at present believed to be the major contribution to thermal resistance at high temperatures.

The mechanisms involved in the intrinsic thermal resistance have been discussed in section 4.5. The most important are undoubtedly those processes (normal and umklapp) which involve three acoustic phonons. Probably of increasing significance as the temperature rises are three-phonon processes involving optical as well as acoustic phonons, and four-phonon processes. A basic study of the effects of the three-acoustic-phonon processes was made by Leibfried and Schlömann (1954). Their calculation was based on a simplified variational method with the strength of the anharmonic interaction being specified by the Grüneisen parameter γ. The resulting formula is

$$\lambda_{ph} = \tfrac{24}{5}4^{1/3}\left(\frac{k_B}{h}\right)^3\frac{M\delta\Theta_D^3}{(\gamma+\frac{1}{2})^2T}, \tag{5.11}$$

where M is the mean atomic mass and δ is the cube root of the atomic volume[(1)]. Leibfried and Schlömann's original formula contained γ^2 instead

[(1)] The appropriate value of the numerical constant in equation (5.11) is a matter of some controversy. The constant given by Leibfried and Schlömann was, in fact, one half of that given above, but Julian (1965) and Hamilton and Parrott (1969) have shown that Leibfried and Schlömann overestimated the phonon scattering strength by a factor of two. A calculation by Parrott (1972) using the same simplified variational method but taking account of the effects of the difference between longitudinal and transverse sound velocities gave an answer about 10% higher. On the other hand Roufosse and Klemens (1973) obtained a result smaller by a factor of twenty-seven. They used a relaxation time method which should normally be expected to give about twice as large a result as the variational method, but they combined this with transition probabilities sixteen times greater.

of $(\gamma+\frac{1}{2})^2$ but Steigmeier and Kudman (1966) showed that the latter was more accurate. This equation which makes $\lambda_{ph} \propto T^{-1}$ must be expected to break down progressively as the temperature increases, owing to the growth of acoustical–optical scattering and four-phonon scattering.

A careful comparison of equation (5.11) with experiment has been made by Steigmeier and Kudman (1966). They considered a large number of semiconductors crystallising in the diamond or zinc blende lattices, taking in each case the value of λ_{ph} at the Debye temperature. They found that in all cases the theoretical value was too large but that the magnitude of the discrepancy was a function of the mass ratio of the two atoms in the unit cell as shown in figure 5.13. This they interpreted as evidence of appreciable acoustical–optical phonon scattering, since the strength of this scattering is to be expected to show a maximum at a certain mass ratio and to disappear altogether when the ratio becomes too large. This is in line with a calculation of the effects of optical phonon scattering by Hugon and Veyssie (1965) who obtained for the additional thermal resistivity

$$W_{op} = \frac{9}{32}\frac{(6\pi^2)^{1/3}}{\pi}\left(\frac{h}{k_B}\right)^3\frac{(\gamma+\frac{1}{2})^2 \mathcal{R}}{M\delta\Theta_D^2} \tag{5.12}$$

where

$$\mathcal{R} = \left(\frac{\omega_0^2}{\omega_D^2}-1\right)\left(6\frac{\omega_0}{\omega_D}-4-\frac{\omega_0^3}{\omega_D^3}\right)\Bigg/\left[1-\exp\left(-\frac{\hbar\omega_0}{k_BT}\right)\right].$$

Here ω_0 is the average frequency of the optical phonons and ω_0/ω_D will be a function of the mass ratio.

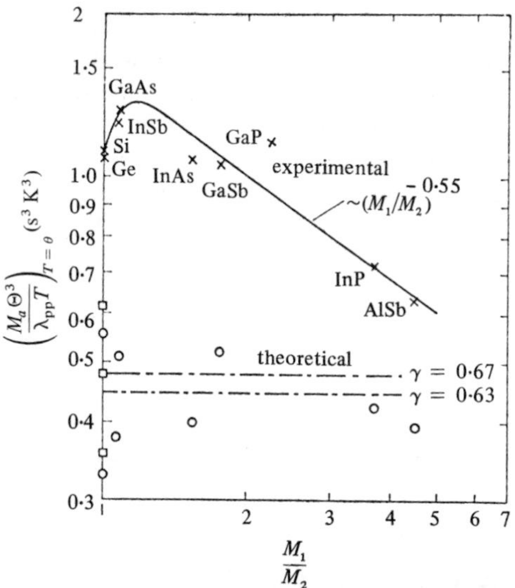

Figure 5.13. Effect of mass ratio of the constituent elements on the reduced thermal resistivity of diamond and zinc blende semiconductors. [Steigmeier and Kudman, 1966, figure 6.]

An alternative treatment of the high temperature lattice thermal conductivity is to use methods similar to those employed at low temperatures and described in section 5.3. A recent example of this is the analysis of the data obtained by Fulkerson *et al.* (1968) on silicon by Joshi and Verma (1970). This gives (i) separate treatment of transverse and longitudinal phonons, (ii) allowance for the dispersion of the phonon frequencies, and (iii) includes three- and four-acoustic-phonon scattering. In this way these workers were able to reproduce the experimental curve, in particular a definite change in slope which had been observed. Two factors, however, diminish the value of this kind of study. Firstly the number of adjustable parameters is quite large, and secondly the actual values of these parameters are not related to any other physical parameter of the material, such as the Grüneisen constant. This means that its predictive usefulness is negligible.

The theory of thermal conductivity shows to better advantage in the analysis of the results obtained from solid solutions of semi-conductors. An important example is the series of germanium–silicon alloys studied by Abeles *et al.* (1962). Theoretically such alloys are regarded as a perfect virtual crystal to which strong atomic disorder is added. The virtual crystal shows properties which are an appropriate average of the properties of the two constituents; in the case of thermal conductivity the Leibfried–Schlömann formula is used with average values of M and Θ_D. The Grüneisen constant would be regarded as approximately constant over the range of germanium–silicon alloys. The atomic disorder is then taken into account by using equation (4.24) possibly with the addition of strain field effects etc., to the mass-difference scattering which that equation already includes. It is then possible using some of the formulae in section 4.6 to calculate $\lambda_{ph}/\lambda_{ph0}$ where λ_{ph0} is the lattice thermal conductivity of the virtual crystal. The simplest way of doing this is to use the Debye formula (4.32) in its high temperature form without considering normal processes at all. This was done by Klemens (1960) who used relaxation times of the following form:

(i) umklapp scattering

$$\tau_U^{-1} = C_U T\omega^2,$$

(ii) disorder scattering

$$\tau_D^{-1} = \mathscr{A}\omega^4,$$

where $\mathscr{A} = \Gamma/4\pi N v^3$ as in equation (4.25). His result was that

$$\frac{\lambda_{ph}}{\lambda_{ph0}} = \frac{\omega_0}{\omega_D}\tan^{-1}\left(\frac{\omega_D}{\omega_0}\right), \tag{5.13}$$

where ω_D is the Debye frequency and

$$\left(\frac{\omega_D}{\omega_0}\right)^2 = \frac{2\pi^2 C\lambda_{ph0}\omega_D\mathscr{A}}{k_B} = \frac{\pi\Gamma\lambda_{ph0}\Theta_D}{2N\hbar v^2};$$

ω_0 is, in fact, the frequency at which $\tau_U = \tau_D$. Unfortunately this simple formula was not in good agreement with experiment, because it neglected the N-processes. If we use the Callaway formula (4.37) and add to the umklapp and disorder scattering an N-process term

$$\tau_N^{-1} = C_N T\omega^2 = k_1 C_U T\omega^2 ,$$

so that k_1 measures the relative strength of N- and U-processes, then a more complicated equation may be derived:

$$\frac{\lambda_{ph}}{\lambda_{ph0}} = \frac{1}{1+\frac{5}{9}k_1}\left\{\frac{1}{y}\tan^{-1}y + \frac{[1-(\tan^{-1}y)/y]^2}{[(1+k_1)/k_1](\frac{1}{5}y^4)-\frac{1}{3}y^3+1-(\tan^{-1}y)/y}\right\}, \tag{5.14}$$

where

$$y^2 = \frac{(\omega_D/\omega_0)^2}{1+\frac{5}{9}k_1}$$

(Abeles, 1963; Parrott, 1963). This equation is plotted in figure 5.14 which shows clearly that the effect of the N-processes is to increase the scattering power of the atomic disorder in the manner described in general terms in section 4.5. The experimental data for germanium–silicon alloys are also shown and may be seen to agree quite well with the theory. The only remaining question is whether an improved fit is provided when a different frequency dependence of the N-process scattering relaxation time, $\tau_N^{-1} \propto T\omega$, is assumed, as proposed by Parrott.

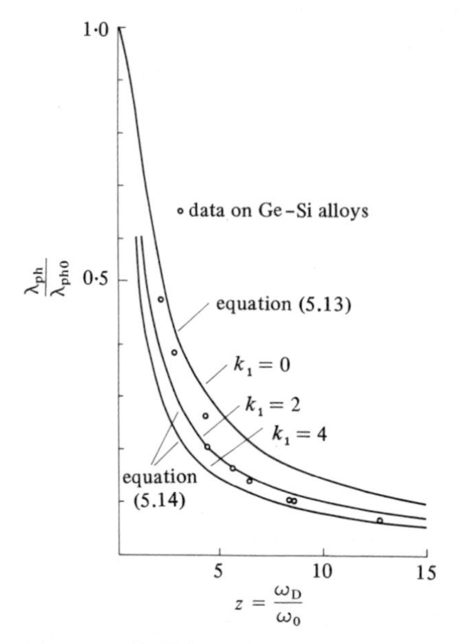

Figure 5.14. Effect of normal processes on atomic disorder scattering. [Parrott, 1963, figure 1.]

The next point to be considered will be the scattering of phonons by free electrons. This has been discussed elsewhere for the case of metallic alloys, where it was found that the effect was limited to rather low temperatures, and this in cases where the electron concentrations are very high. Since even the most heavily doped semiconductors have electron concentrations two or three orders of magnitude lower, it might be supposed that scattering of phonons by electrons could be neglected in semiconductors at high temperatures. That this is not the case for materials showing very strong point-defect scattering was pointed out by Steigmeier and Abeles (1964). At high temperatures the integrand in the thermal conductivity integral, equation (4.32), is proportional to τz^2, where τ is the total relaxation time and $z = \omega/\omega_D$. In figure 5.15 τz^2 is plotted as a function of z for different combinations of scattering mechanisms. The important point to notice is that electron scattering has a far greater relative effect when strong point-defect scattering is added to the three-phonon scattering. This is because point defects scatter strongly at high frequencies, while electrons scatter more at low frequencies. Thus the same electron concentration which would have a negligible effect on the thermal conductivity of pure silicon would have a strong effect in a germanium–silicon alloy. These ideas were applied to heavily doped germanium–silicon alloys by Steigmeier and Abeles.

A somewhat similar situation prevails in the case of boundary scattering, such as might occur in a fine grained polycrystal. This effect is also enhanced in materials where disorder scattering is very strong as was pointed out by Goldsmid and Penn (1968). The consequences of this have been further explored by Parrott (1969) and experimental confirmation obtained by Savvides and Goldsmid (1973) who used silicon disordered by neutron irradiation.

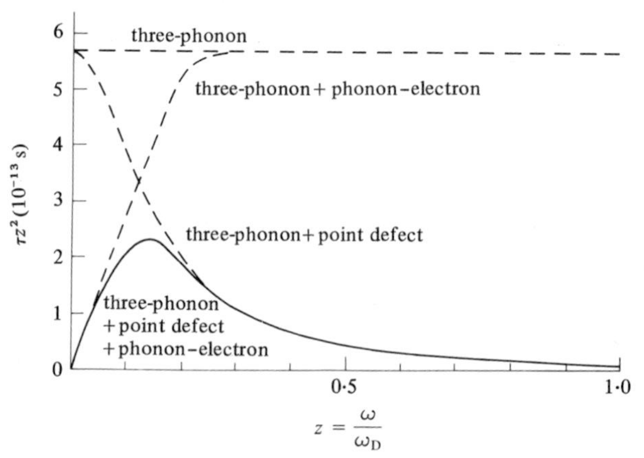

Figure 5.15. Thermal conductivity integrand for different scattering processes. [Steigmeier and Abeles, 1964, figure 4.]

This completes the discussion of the analysis of high-temperature thermal conductivity of insulators and semiconductors.

5.8 The prediction of thermal conductivity

In many ways, the ability to predict is the most important requirement of physical theory. It is therefore necessary to consider what can be done in the case of thermal conductivity. Before going on, however, there are two points which should be borne in mind. The first of these concerns the question of how accurate a prediction is necessary. This will of course vary widely according to the application for which the prediction is needed. The second concerns the information on which the prediction is to be based. Clearly there is not much practical value in a prediction which requires data considerably less accessible to measurement than the thermal conductivity.

It will be clear that a preliminary assessment of the likely relative magnitude of the electronic and lattice thermal conductivity should be made. There is also the question as to whether any other heat carrying mechanism may be at work, but this is sufficiently unlikely that the possibility can be ignored in practice. In dealing with the electronic heat conduction little can be done without some information about the electrical conductivity, but this is usually readily available, if only at room temperature. If this shows that the electrical conduction is of metallic type, then it is of great assistance to know that the electronic thermal conductivity above room temperature will generally not vary much with temperature. The difficulty arises at low temperatures. Here the ideal thermal resistance will be satisfactorily given by White and Woods' empirical relation, equation (5.3), but at sufficiently low temperatures this will be dominated by the residual resistance which varies considerably between different samples of the same material. This means that it is necessary to have a measurement of the electrical conductivity at fairly low temperatures. Once this is available, a quite accurate estimate of the electronic thermal conductivity can be made by the methods discussed in section 5.2 of this chapter. Unless one is dealing with a high-purity metal, however, it will be necessary to estimate the lattice conduction, a point to which we return later.

In most semiconductors the contribution of electrons or holes to the thermal conductivity is quite negligible. Nevertheless, if the material is either very heavily doped or intrinsic, it may be necessary to allow for the heat transported by the free carriers. Where this is the case the methods of section 5.7 can normally be used with success. Since it is unusual even in these cases, for the lattice conduction not to be a major proportion of the total, rather large percentage errors are usually acceptable.

The main problem of prediction undoubtedly arises in the case of the lattice heat conduction. We consider first the high-temperature situation where phonon–phonon scattering dominates the thermal resistance.

In essence all discussion of this problem goes back to the Leibfried–Schlömann formula. This means immediately that fairly large relative errors are likely, particularly as this formula does not give the temperature dependence very accurately. The major difficulty with using this formula is that some of the quantities involved are not at all accessible, in particular the Grüneisen constant and the Debye temperature. The latter poses the major problem. One way of tackling it, along lines suggested originally by Keyes (1959), involves using the Lindemann melting formula (Zemansky, 1968). This suggests the relation

$$\frac{M\delta^2\Theta_D^2}{T_M} = \mathscr{A},$$

where $\mathscr{A}$ should be constant for materials of a given type. Alternatively, Steigmeier (1963) has suggested a relation of the form

$$M\delta^3\Theta_D^2 = \mathscr{B},$$

where, again, $\mathscr{B}$ should be constant for materials of a similar type. The Grüneisen constant varies much less than Θ_D and so does not pose so much difficulty. Indeed, Steigmeier's formula suggests that $\gamma = \frac{1}{2}$, which is fairly accurate for tetrahedrally bonded semiconductors. Assistance can also be derived from the paper of Steigmeier and Kudman (1966) referred to earlier and from recent work by Spitzer (1970); this latter contains a very comprehensive collection of experimental data. The effect of having a large and complex unit cell has been considered by Roufosse and Klemens (1973) who conclude that the conductivity should vary as the inverse cube root of the number of atoms in the unit cell. In this way the lattice thermal conductivity in the neighbourhood of the Debye temperature can probably be estimated to within about 50%. It will be necessary to consider such a term, not only in semiconductors and insulators, but also for those metals and alloys whose electrical conductivity is low. The effects of alloying in semiconductors, etc., can be predicted quite well with equation (5.14).

At low temperatures it is difficult to forecast lattice thermal conductivity except in the boundary scattering regime. There are two main reasons for this. Firstly the thermal conductivity at and around its maximum is largely limited by lattice defect scattering, which will vary widely from one sample to another. Secondly the three-phonon scattering at low temperatures (i.e. between the temperature of the maximum and about $\frac{1}{2}\Theta_D$) is difficult to determine even when its strength is known at, say, room temperature. There is no equation for the relaxation time which enables one to interpolate between high and low temperature behaviour. Errors of a factor of two or more are to be anticipated in the neighbourhood of the maximum.

References

Abeles, B., Beers, D. S., Cody, G. D., Dismukes, J. P., 1962, *Phys. Rev.,* **125**, 44.
Abeles, B., 1963, *Phys. Rev.,* **131**, 1906.
Agrawal, B. K., 1967, *Phys. Rev.,* **162**, 731.
Amith, A., Kudman, I., Steigmeier, E., 1964, *Phys. Rev.,* **136**, A1149.
Archibald, M. A., Dunick, J. E., Jericho, M. H., 1967, *Phys. Rev.,* **153**, 786.
Armitage, D., Goldsmid, H. J., 1969, *J. Phys. C,* **2**, 2138.
van Baarle, C., Gorter, F. W., Winsermuis, P., 1967, *Physica,* **35**, 223.
Bäcklund, N. G., 1961, *J. Phys. Chem. Solids,* **20**, 1.
Beers, D. S., Cody, G. D., Abeles, B., 1962, *Proceedings of the International Conference on the Physics of Semiconductors* (Institute of Physics and the Physical Society, London), p.41.
Berman, R., 1951, *Proc. Roy. Soc. A,* **208**, 90.
Berman, R., Brock, J. C. F., 1965, *Proc. Roy. Soc. A,* **289**, 46.
Bhandari, C. M., Verma, G. S., 1965, *Phys. Rev.,* **140**, A2101.
Brown, M. A., Popovic, Z., 1972, *J. Phys. C,* **5**, 2317.
Callaway, J. C., 1959, *Phys. Rev.,* **113**, 1046.
Charsley, P., Salter, J. A. M., Leaver, D. W., 1968, *Phys. Stat. Solidi,* **25**, 531.
Choy, C. L., Salinger, G. L., Chiang, Y. C., 1970, *J. Appl. Phys.,* **41**, 597.
Dreyfus, B., Fernandes, N. C., Maynard, R., 1968, *Phys. Letts.,* **26A**, 647.
Farrell, T., Grieg, D., 1969, *J. Phys. C,* **2**, 1465.
Fulkerson, W., Moore, J. P., Williams, R. K., Graves, R. S., McElroy, D. L., 1968, *Phys. Rev.,* **167**, 765.
Geballe, T. H., Hull, G. W., 1958, *Phys. Rev.,* **110**, 773.
Goff, J. F., 1970, *Phys. Rev.,* **B1**, 1351.
Goldsmid, H. J., Penn, A. W., 1968, *Phys. Lett.,* **27A**, 523.
Hamilton, R. A. H., Parrott, J. E., 1969, *Phys. Rev.,* **178**, 1284.
Harrington, J. A., Walker, C. T., 1970, *Phys. Rev.,* **B1**, 882.
Herring, C., 1967, *Phys. Rev. Letts.,* **19**, 167 (see Erratum also).
Holland, M. G., 1963, *Phys. Rev.,* **132**, 2461.
Hugon, P. L., Veyssie, J. J., 1965, *Phys. Stat. Solidi,* **8**, 561.
Hurst, W. S., Frankl, D. R., 1969, *Phys. Rev.,* **186**, 801.
Joshi, Y. P., Verma, G. S., 1970, *Phys. Rev.,* **B1**, 750.
Julian, C. L., 1965, *Phys. Rev.,* **137**, A128.
Kadanoff, L. P., Martin, P. C., 1961, *Phys. Rev.,* **124**, 670.
Kelly, B. T., 1973, *High Temperatures - High Pressures,* **5**, 133.
Keyes, R. W., 1959, *Phys. Rev.,* **115**, 564.
Klemens, P. G., 1960, *Phys. Rev.,* **119**, 507.
Korenblit, L. L., Sherstobitov, V. E., 1968, *Fiz. Tekh. Poluprov.,* **2**, 688.
Leibfried, G., Schlömann, E., 1954, *Nachr. Akad. Wiss. Göttingen; Math. Phys. Kl.,* **4**, 71.
Lindenfeld, P., Pennebaker, W. B., 1962, *Phys. Rev.,* **127**, 188.
Lindenfeld, P., Lynton, E. A., McLachlan, D. S., Souler, R., 1966, *Phys. Rev.,* **143**, 434.
Lomer, J. N., Rosenberg, H. M., 1959, *Phil. Mag.,* **4**, 467.
Lynton, E. A., 1964, *Superconductivity* (Methuen, London).
Mendelssohn, K., Shiffman, C. A., 1959, *Proc. Roy. Soc. A,* **255**, 199.
Mendelssohn, K., Rosenberg, H. M., 1961, *Solid State Physics,* **12**, 233.
Mezhov-Deglin, L. P., 1965, *Zh. Eksper. Teor. Fiz.,* **49**, 66.
Moss, M., 1966, *J. Appl. Phys.,* **37**, 4168.
Nellis, W. J., Levgold, S., 1969, *Phys. Rev.,* **180**, 581.
Parrott, J. E., 1963, *Proc. Phys. Soc.,* **81**, 726.
Parrott, J. E., 1969, *J. Phys. C,* **2**, 147.

Parrott, J. E., 1972, International Conference on Phonon Scattering in Solids, Paris, France (Documentation Service of CEN Saclay, France), p.27.
Pippard, A. B., 1955, *Phil. Mag.,* **46**, 1104.
Putley, E. H., 1960, *The Hall Effect and Related Phenomena* (Butterworths, London).
Rickaysen, G., 1968, *J. Phys. C,* **1**, 744.
Rosenberg, H. M., 1955, *Phil. Trans. Roy. Soc. A,* **247**, 441.
Roufosse, M., Klemens, P. G., 1973, *Phys. Rev.,* **B7**, 5379.
Satterthwaite, C. B., 1962, *Phys. Rev.,* **125**, 873.
Savvides, N., Goldsmid, H. J., 1973, *J. Phys. C.,* **6**, 1701.
Schriempf, J. T., 1968, *Phys. Rev. Letts.,* **20**, 1034.
Schwartz, J. W., Walker, C. T., 1967, *Phys. Rev.,* **155**, 959.
Seward, W. D., Lazarus, D., Fain, S. C., 1969, *Phys. Rev.,* **178**, 345.
Slack, G. A., Glassbrenner, C., 1960, *Phys. Rev.,* **120**, 782.
Sousa, J. B., 1968, *Phys. Lett.,* **26A**, 607.
Sousa, J. B., 1969, *J. Phys. C,* **2**, 629.
Spitzer, D. P., 1970, *J. Phys. Chem. Solids,* **31**, 19.
Srivastava, B. N., Chatterjee, S., Sen, K., Chakraborty, D. K., 1970, *J. Phys. C (Metal Phys. Suppl.),* **35**, 169.
Stauder, B. F., Mielczarek, E. V., 1967, *Phys. Rev.,* **158**, 630.
Steigmeier, E. F., 1963, *Appl. Phys. Letts.,* **3**, 6.
Steigmeier, E. F., Abeles, B., 1964, *Phys. Rev.,* **136**, A1149.
Steigmeier, E. F., Kudman, I., 1966, *Phys. Rev.,* **141**, 767.
Taylor, R., 1966, *Phil. Mag.,* **13**, 157.
White, G. K., Woods, S. B., 1958, *Phil. Mag.,* **3**, 342.
White, G. K., Woods, S. B., 1959, *Phil. Trans. Roy. Soc. A,* **251**, 273.
White, G. K., 1960, *Austral. J. Phys.,* **13**, 255.
Worlock, J. M., 1966, *Phys. Rev.,* **147**, 636.
Zemansky, M. W., 1968, *Heat and Thermodynamics*, 5th edition (McGraw-Hill, New York), p.376.

The thermal conductivity of technological materials

6.1 Introduction

Of the materials of technological importance, the behaviour of metals, alloys, and semiconductors has been adequately discussed in chapters 4 and 5. However, many solids in which lattice conduction predominates are far from ideal—they are usually polycrystalline, often impure, and may consist of two or more components, one of which is frequently air.

Figure 6.1 indicates the thermal conductivity of a number of different solids above 200 K. The highest value shown is that of silver at about 5 W cm^{-1} K^{-1} (only diamond and pyrolytic graphite have higher thermal conductivities). The lowest known values are $\sim 10^{-4}$ W cm^{-1} K^{-1}, for microporous solids. In general, the single phase crystalline solids are all included in the top half of the range, while the porous solids such as firebrick, and the disordered materials such as glass and polymers all occur in the lower half.

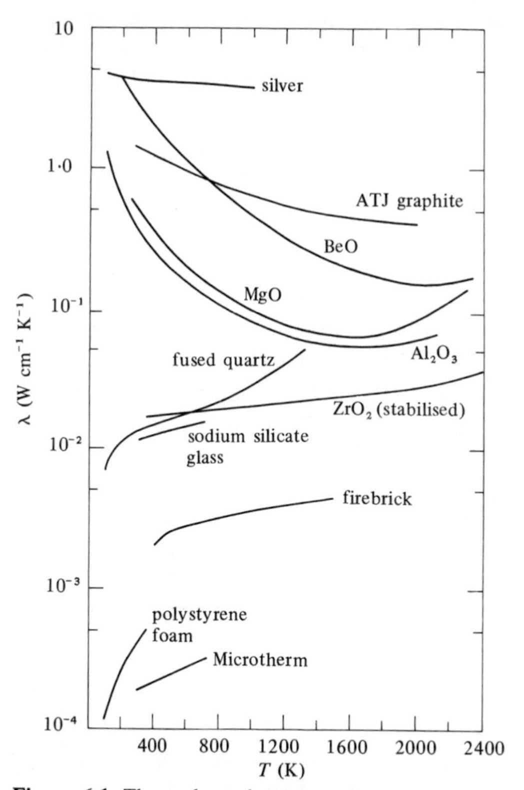

Figure 6.1. Thermal conductivity of various solids.

6.2 Single-phase crystalline solids

6.2.1 Isotropic oxides

The thermal conductivities of several oxides of importance in the ceramics industry are included in figure 6.1. The magnitudes are widely different at 300 K and show a systematic decrease with increasing atomic weight of the metal ion. This is due to an increasing anharmonicity in the lattice vibrations as the atomic weight of the cation becomes increasingly large compared with that of the oxygen anion.

It can be seen from figure 6.1 that above 300 K the thermal conductivity of all the oxides, with the exception of zirconia, decreases with increasing temperature and above 1000 K there is a much smaller spread in values. In some cases there is an increase in thermal conductivity at the highest temperatures due to radiative heat transfer. Approximate values of the Debye temperature of these oxides is shown in table 6.1.

Above 300 K it would therefore be expected that scattering by umklapp processes would be starting to be important but only UO_2 would be expected to show, in the vicinity of 300 K, the predominance of umklapp scattering and the consequent T^{-1} dependence. In figure 6.2 the thermal conductivities of these oxides are plotted as functions of $1/T$ and there is a linear portion in each case over a limited range. The curves indicate, however, that it is unwise to assume this dependence to hold over too wide a range of temperature.

At the highest temperatures (neglecting radiative heat transfer) the mean free path of the phonons is ultimately limited by interatomic spacing, thus reducing the spread in thermal conductivity of different crystalline solids.

When umklapp scattering predominates in isotropic solids there should be little difference between the thermal conductivity of single-crystal and polycrystalline samples. Results for both single-crystal and polycrystalline MgO and Al_2O_3 are given in figure 6.2. Owing to the spread in published thermal conductivity data, the smoothed values recommended by the Thermophysical Research Centre at Purdue (Touloukian, 1970) have been used. In each case the values for the single crystal are higher than those given for the polycrystalline sample of 98% theoretical density and 99·5% purity. The difference is greater than that

Table 6.1. Debye temperature of some refractory oxides.

Oxide	Θ_D (K)
BeO	900
Al_2O_3	970
MgO	785
TiO_2	760
UO_2	377

given by a correction for 2% porosity made as shown in section 6.3.2 using equation (6.11). Since a reliance of 6% on the polycrystalline data is recommended and of 10–15% on the single crystal data, it is pointless to comment on the magnitude of the difference.

The low thermal conductivity of zirconia is in keeping with the complex structure of its stabilised form and illustrates the greater tendency of the more complicated lattice structures to scatter phonons.

In general grain boundary scattering has most effect on the maximum of the thermal conductivity curve which generally occurs well below the Debye temperature. However, as the grain size becomes smaller the peak in the conductivity occurs at increasingly high temperatures, the maximum conductivity decreases and the peak in the curve becomes much less sharp. Thus the curve for UO_2 would be reasonable for polycrystalline material of very small grain size. It is therefore surprising that Moore and McElroy (1971) claim that this curve is representative of both single crystal and polycrystalline UO_2.

Small quantities of impurities lower the thermal conductivity, the effect being greatest in the region of the peak in the thermal conductivity–temperature curve. At room temperature the increase in thermal resistance is usually small and remains constant as the temperature is increased. However, the addition of a second component which forms a solid solution greatly decreases the thermal conductivity, often to a value

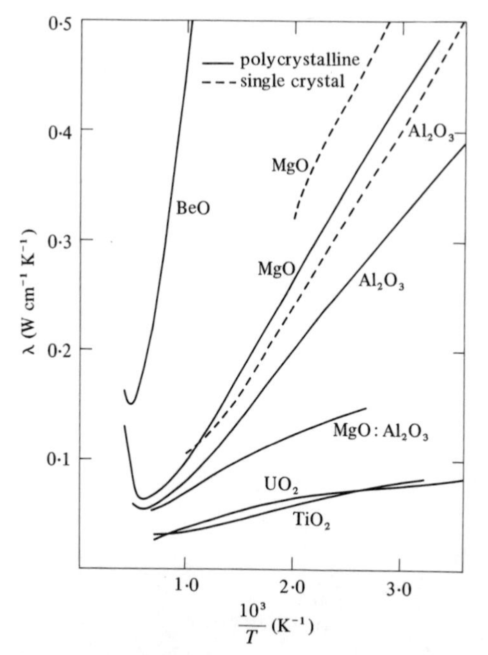

Figure 6.2. Thermal conductivity of some refractory oxides.

less than that of either component. This is clearly illustrated by the MgO–NiO system (Kingery, 1959) as shown in figure 6.3. Many similar systems in the field of semiconductors, such as Ge–Si (see section 5.7), UO_2–ThO_2, Sb_2Te_3 and Bi_2Se_3 in Bi_2Te_3, and PbSe in PbTe have been investigated. Such solid solutions disturb the short range order of the lattice, and consequently the phonons responsible for heat conduction, with wavelengths of the order of a few interatomic spacings, are effectively scattered by such disturbances. Another example in common use is spinel, magnesium aluminate $MgAl_2O_4$, a compound of MgO and Ai_2O_3; as shown in figure 6.2 this has a considerably lower conductivity than that of either of its constituents despite their similar crystal structure.

An interesting effect was observed by Cosgrove *et al.* (1961) in crystals of the semiconducting alloy $BiSbTe_3$ doped with Se, grown under different conditions. A decrease in the apparent lattice thermal conductivity (estimated by subtracting the calculated electronic component from the measured value) was found with decreasing freezing rate and with increasing temperature gradient at the liquid/solid interface. At the same time the Seebeck coefficient and the electrical conductivity increased. This was attributed to a variation in the degree of microsegregation produced in the crystals as the freezing conditions were changed. Ure (1962) assumed that electric currents were circulating as a result of

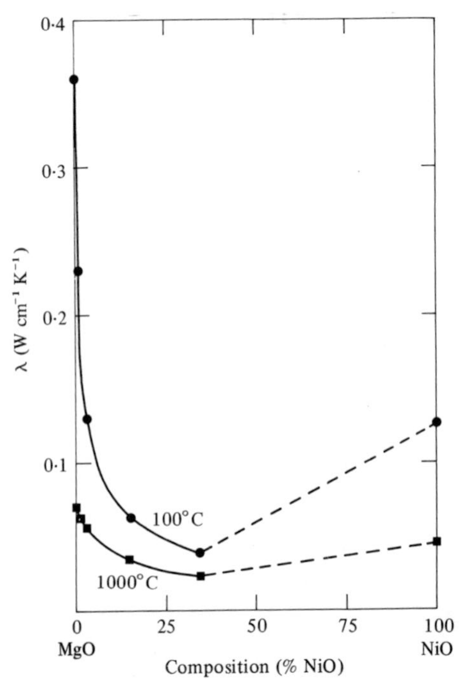

Figure 6.3. Thermal conductivity of the MgO–NiO system.

thermoelectric effects arising from different electrical properties of the two phases and that additional heat was transported by the circulating currents. He was able to account for the maximum observed differences in apparent lattice conductivity of a factor two by assuming a 33% segregation. Such effects will of course only be appreciable in good thermoelectric materials, but it shows the importance of crystal growth conditions for devices whose performance is related to the thermoelectric figure of merit.

6.2.2 Anisotropic solids

Depending upon crystal structure, the thermal conductivity of some refractory solids is anisotropic with its greatest value in the direction of the greatest bonding forces. Graphite with its hexagonal structure shows the greatest known anisotropy, with the thermal conductivity in the *a* direction (parallel to the basal plane) some two to three hundred times greater than that in the *c* direction (parallel to the hexagonal axis). Although small single crystals have been obtained from natural sources, more massive samples of highly oriented pyrolytic graphites are of more practical significance. The thermal conductivity of these, particularly below 300 K, depends upon the crystallite size. Typical values for a pyrolytic graphite, quoted by TPRC (Touloukian, 1970) are shown in figure 6.4. In the *a* direction it is a better conductor than most refractory

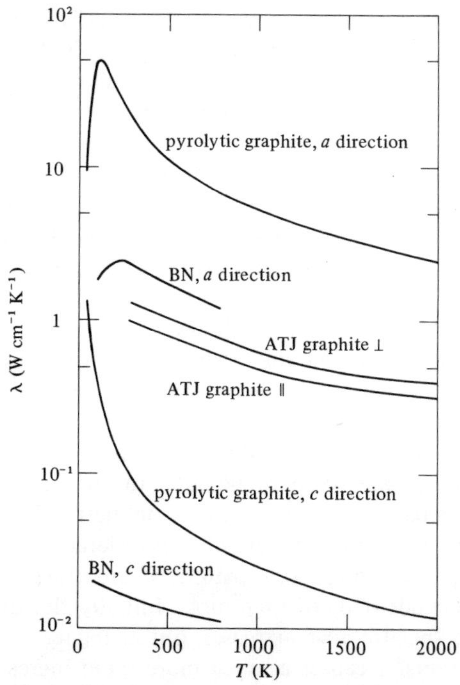

Figure 6.4. Thermal conductivity of graphite and pyrolytic boron nitride.

materials while in the c direction it has one of the lowest values for a crystalline solid, comparable with that of UO_2. Recommended values for polycrystalline graphite made from pitch-bonded petroleum coke, parallel and perpendicular to the direction of pressing, are also shown in figure 6.4 and lie intermediate to those in the a and c directions. It has been shown by Kelly (1969) that the principal conductivities of a polycrystalline graphite with cylindrical symmetry can be expressed in terms of an orientation function $I(\psi)$ which represents the relative density of basal plane normals per unit solid angle about the angle of inclination ψ relative to the symmetry axis of the distribution. In many cases it is possible to express $I(\psi)$ as

$$I(\psi) = \cos^m\theta; \tag{6.1}$$

then

$$\lambda_{\perp} = \lambda_a\left[1-\left(1-\frac{\lambda_c}{\lambda_a}\right)\left(\frac{1}{m+3}\right)\right] \tag{6.2}$$

$$\lambda_{\parallel} = \lambda_a\left[1-\left(1-\frac{\lambda_c}{\lambda_a}\right)\left(\frac{m+1}{m+3}\right)\right] \tag{6.3}$$

where λ_a and λ_c are appropriate to the crystallite size of the polycrystalline material. The orientation function can be evaluated from the integrated intensities of the x-ray reflections from the (002) planes. However, it is found that the experimentally determined conductivities in polycrystalline graphites are always lower than those predicted by equations (6.2) and (6.3), even when allowance has been made for porosity as shown in section 6.3.2. It is necessary to include an empirical tortuosity factor to allow for the effects of pore shape and size.

A material similar in structure to graphite is boron nitride. This also shows a marked anisotropy (Simpson and Stuckes, 1971), although not as great as that of graphite, as can be seen from figure 6.4, but this could be due to the fact that material as well oriented as pyrolytic graphite has not yet been made.

Another technological material showing a small degree of anisotropy is quartz. Its thermal conductivities parallel and perpendicular to the c axis are shown in figure 6.5; both decrease with increasing temperature as expected for a crystalline solid. For comparison, values for fused quartz are shown in the same figure. Its thermal conductivity, up to 800 K at least, is lower than that either parallel or perpendicular to the c axis in the single crystal. The comparatively low thermal conductivity, increasing with increasing temperature is typical of a disordered or amorphous solid where the phonon mean free path is of the order of the interatomic spacing and independent of temperature. Initially the thermal conductivity increases as the specific heat increases, but at higher temperatures radiative heat transfer causes a much more rapid increase in apparent thermal conductivity.

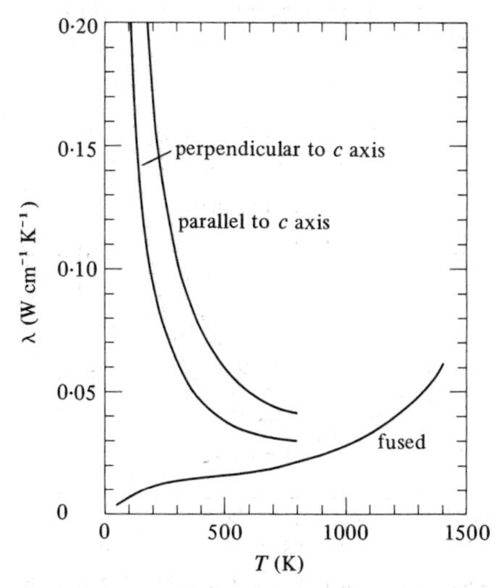

Figure 6.5. Thermal conductivity of quartz.

6.3 Heterogeneous solids

6.3.1 Theoretical models for the thermal conductivity of mixtures

The thermal conductivity of single-phase materials is fairly well understood but most materials in everyday use are heterogeneous. Even polycrystalline solids have a density below that of the single crystal owing to the presence of voids within the lattice. Perhaps more obvious examples are concrete, bricks, wood, ceramics, and fibre-reinforced materials, all of which have at least two components, and in many of their uses thermal conductivity is an important parameter. In general, the thermal behaviour of the heterogeneous material, unlike that of solid solutions, lies between that of its components and depends upon the volume of each and its distribution. It is frequently desirable to estimate the properties of composite materials from a knowledge of the properties of the individual components. It is necessary to choose a model which represents the distribution of the components within the mixture.

The simplest possible arrangement of a mixture of two components is with the materials arranged in parallel slabs as shown in figure 6.6. The maximum conductivity of such an arrangement, λ_{mix}, occurs when the heat flow is parallel to the plane of the slabs and is given by

$$\lambda_{mix(max)} = \phi_1\lambda_1 + \phi_2\lambda_2 \tag{6.4}$$

where ϕ_1 and ϕ_2 are the volume fractions of the components of conductivity λ_1 and λ_2 respectively, and $\phi_1 + \phi_2 = 1$. Heat conduction in this arrangement is predominantly through the better conductor (compare electrically with parallel resistors) and when $\lambda_1 \gg \lambda_2$ then $\lambda_{mix(max)} \approx \phi_1\lambda_1$.

However, for heat flow perpendicular to the plane of the slabs (equivalent to resistors in series) the heat flow through each component must be equal but the temperature gradient in each is different. The total conductivity is now

$$\lambda_{\mathrm{mix(min)}} = \frac{\lambda_1 \lambda_2}{\phi_1 \lambda_2 + \phi_2 \lambda_1} \, . \tag{6.5}$$

It is dominated by the poorer conductor and is the minimum conductivity of the arrangement. The ratio $\lambda_{\mathrm{mix}}/\lambda_1$ for parallel and series flow is shown by curves *1* and *2* in figure 6.7 for λ_1/λ_2 equal to 10. The general structure of any real material corresponds to a combination of series and parallel arrangements so that the thermal conductivity of any two-component system having $\lambda_1/\lambda_2 = 10$ will lie between these two curves.

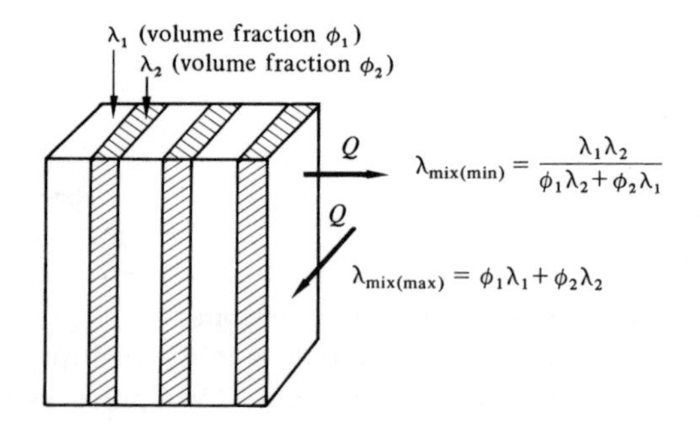

Figure 6.6. Simple parallel-slab model.

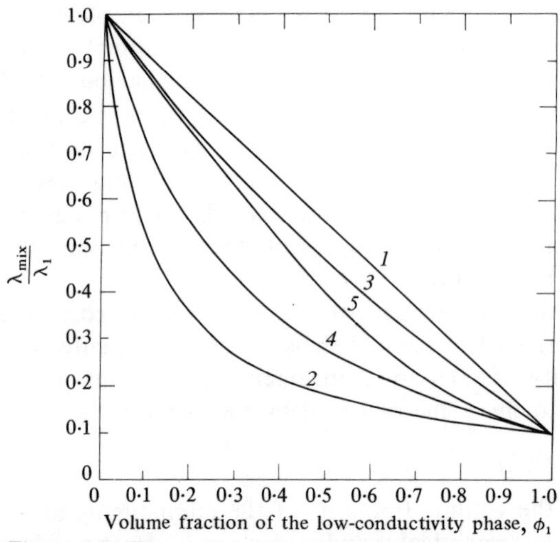

Figure 6.7. Theoretical thermal conductivity of a two-phase medium having $\lambda_1/\lambda_2 = 10$: *1* flow through slabs in parallel; *2* flow through slabs in series; *3* Maxwell model, high-conductivity phase continuous; *4* Maxwell model, low-conductivity phase continuous; *5* Brailsford and Major model.

Various relationships have been suggested to predict more realistically the behaviour of two-component systems. The simplest empirical approach defines the conductivity of the system as the geometric mean of the separate conductivities, that is

$$\lambda_{mix} = \lambda_1^{\phi_1}\lambda_2^{(1-\phi_1)} \tag{6.6a}$$

or

$$\log\lambda_{mix} = \phi_1 \log\lambda_1 + (1-\phi_1)\log\lambda_2 \ . \tag{6.6b}$$

This is mathematically unsound and has been found to tend to overestimate the conductivity of the system.

In a more rigorous approach the problem is to calculate the disturbance to linear heat flow through a uniform medium by the presence of a region of different conductivity. It is worth noting that since the magnetic permeability, dielectric constant, electrical conductivity, and thermal conductivity are all described by the Laplace equation, the solutions for all these properties of a heterogeneous body are mathematically identical. In using this approach it is necessary to define a mathematical model which bears some relationship to the microstructure of the heterogeneous material.

One of the earliest models was that of Maxwell (1904) who solved the problem for randomly-sized spheres of one medium randomly distributed in another medium. The thermal conductivity λ_{mix} of such an array is given by

$$\frac{\lambda_{mix}}{\lambda_{cont}} = \frac{1+2\chi-2\phi(\chi-1)}{1+2\chi+\phi(\chi-1)} \tag{6.7}$$

where χ is the ratio of the conductivities of the continuous and the dispersed phases $\lambda_{cont}/\lambda_{disp}$, and ϕ is the fractional volume of the dispersed phase. The conductivity λ_{mix} of a mixture of two phases of conductivity λ_1 and λ_2 where $\lambda_1/\lambda_2 = 10$, is shown in figure 6.7. The ratio λ_{mix}/λ_1 is plotted as a function of the volume fraction of the low-conductivity phase; curve *3* represents the case when the high-conductivity phase, λ_1, is continuous and curve *4* that with the low-conductivity phase, λ_2, continuous. These curves indicate that to increase the conductivity of a poor conductor a greater proportion of a good conductor is required than the proportion of a poor conductor necessary to produce the same percentage decrease in the conductivity of a good conductor.

Since it is a condition of the model that the spheres are sufficiently far apart to have no influence on each other, equation (6.7) will only apply over a limited range, probably up to about 30% dispersed phase. Therefore, when considering the full range of possible combinations, one medium will be continuous at one end of the range and effectively discontinuous at the other end. Thus its conductivity should not follow either curve *3* or *4* continuously but should tend from the left hand side of curve *3* to the right hand side of curve *4*, crossing from one to the other somewhere in the middle of the range. Such S-shaped behaviour has been found by Kingery (1959) in the two phase system $MgO-Mg_2SiO_4$.

Brailsford and Major (1964) have extended the results of the Maxwell model to cover the full range of composition, by regarding a random two-phase assembly composed of the two single phases in the correct proportions, embedded in a random mixture of the same two phases having a conductivity equal to the average value of the conductivity of the two-phase assembly. This leads to a value for the conductivity of the assembly given by

$$\frac{\lambda_{\text{mix}}}{\lambda_2} = \tfrac{1}{4}[(3\phi_1 - 1)\chi + 2 - 3\phi_1 + \{[(3\phi_1 - 1)\chi + 2 - 3\phi_1]^2 + 8\chi\}^{1/2}] \qquad (6.8)$$

where χ is now the ratio λ_1/λ_2; this is shown by curve *5* in figure 6.7 for $\chi = \lambda_1/\lambda_2 = 10$. This model, as shown by Brailsford and Major, can be extended to predict the thermal conductivity of a system containing three or more phases.

Specimens specially constructed to conform with the two-phase assembly of Maxwell were found by Sugawara and Yoshizawa (1961) to have thermal conductivities in good agreement with those predicted by equation (6.7). However, in general only approximate predictions can be made on the basis of equations (6.7) or (6.8), as the structure of two-phase systems does not necessarily fit the model. It is found that the Maxwell model agrees with experimental measurements for dispersions of a good conductor in a poor conductor rather better than for dispersions of a poor conductor in a good conductor; in the latter case the dispersed phase tends to have a greater effect than the theory predicts.

There have been several extensions of this type of model, many based upon the original Maxwell approach but allowing the dispersed phase to have some shape other than spherical. Generally either prolate or oblate spheroids are considered (that is approximately disc-shaped or cylindrical inclusions) and the axial ratio of the spheroids is included as a variable. The various formulae have been summarised by Reynolds and Hough (1957) who tabulated the dielectric constant of a two-component system when such spheroidal particles are either randomly dispersed or oriented. Consequently, if the microstructure of a two component system is known, the appropriate equation can be chosen to predict the thermal conductivity by replacing the dielectric constant in the tabulated equations by the appropriate thermal conductivity.

A different type of approach to the problem of the heterogeneous mixture is based on an effective series-parallel model. In a technique developed by Cheng and Vachon (1969), a parabolic distribution of the discontinuous phase is assumed from which it is shown that

$$\frac{1}{\lambda_{\text{mix}}} = 2\int_0^{B/2} \frac{1}{[\lambda_{\text{cont}} + B(\lambda_{\text{disp}} - \lambda_{\text{cont}})] - C(\lambda_{\text{disp}} - \lambda_{\text{cont}})x^2}\,dx + \frac{1-B}{\lambda_{\text{cont}}}, \qquad (6.9)$$

where $B = (\frac{3}{2}\phi)^{1/2}$ and $C = 4(\frac{2}{3}\phi)^{1/2}$; this relation holds for $\phi \leqslant 0{\cdot}667$. When it is used for $\lambda_1/\lambda_2 = 10$, as before, with the better conductor continuous, the results are similar to those of Maxwell shown in curve *3*. However, for the poorer conductor continuous the results agree more closely with those of Brailsford and Major, curve *5*.

6.3.2 Porous solids

Many heterogeneous solids are porous and one component is air or other fluid within the pores. Frequently the thermal conductivity of the fluid is negligible compared with that of the solid; under these conditions equation 6.7 becomes

$$\frac{\lambda_{mix}}{\lambda_{cont}} = \frac{1-\phi}{1+\frac{1}{2}\phi}. \tag{6.10}$$

For low values of ϕ, that is of porosity, this equation can be written

$$\lambda_{mix} \approx \lambda_{cont}(1-\tfrac{3}{2}\phi). \tag{6.11}$$

It has been found that experimental results can sometimes be fitted rather better to the equation

$$\lambda_{mix} = \lambda_{cont}(1-\beta\phi), \tag{6.12}$$

where β is an adjustable parameter. This is effectively a structural factor allowing for pores of shape other than spherical and various values can be found for it theoretically by using the various equations summarised by Reynolds and Hough. However, since the shape of pores is frequently unknown, equation (6.11) is the one which is commonly used to evaluate the thermal conductivity of a crystalline solid from measurements on slightly porous samples.

Solids with a high degree of porosity are becoming increasingly important in technological developments. For example, design of thermal insulating materials depends upon the heat transfer characteristics of porous media, while knowledge of the thermal properties of natural rocks and soils is required in the design of underground storage depots for liquefied natural gas, and for calculations of the current-carrying capacity of buried cables and of heat losses from underground steam pipes. In addition, the thermal conductivity of masonry, concrete, and other building materials and its variation with moisture content are very important to architects and civil engineers in the endeavour to produce a comfortable human environment. It may be required therefore to predict either the thermal conductivity of a heterogeneous solid or the dependence of its magnitude upon temperature, porosity, and type of fluid within its pores.

When considering a porous solid it must be remembered that heat can be transferred across a pore by convection and by radiation as well as conduction. At room temperature, unless the pores exceed about 3 mm in diameter, convection is negligible and the pores must be even larger for convection to play an appreciable part at higher temperatures. Heat transferred by radiation increases as the pore size is increased, and its effect has been calculated by both Russell (1935) and Loeb (1954). As it has a T^3 dependence, radiation obviously plays an increasingly active role at high temperatures.

In general, engineering solids are aggregates of more than two components, often a crystalline phase together with a glass or liquid and a gas. Consequently accurate predictions are not usually possible but general trends in behaviour can be estimated. The thermal conductivity will decrease with increasing proportions of glass, liquid, and pores, although it will start to increase at very high porosities, probably around 80%, due to the increasing effect of radiative heat transfer. Also, as the proportion of a glassy phase or of porosity within a crystalline solid increases, the temperature coefficient of the thermal conductivity changes from negative to positive, the effect being enhanced at high temperatures owing to increasing radiation.

In order to predict the magnitude of thermal conductivity Maxwell's model is the simplest to apply, particularly if the problem can be reduced to one where only two components are present. If, for example, one has a solid composed of pores and small crystals embedded in a glass, then the solid component can be taken as an isotropic phase of the crystalline solid in glass and its thermal conductivity evaluated by means of equation (6.7). The thermal conductivity of the heterogeneous solid can then be considered as a mixture of this solid phase with air and equation (6.7) used again. For porosities less than 50% the solid component can be considered to be the continuous phase. However, for higher porosities, owing to the restrictions of the model, it is necessary to consider the gas phase to be continuous and this tends to underestimate the conductivity. Under these conditions it is better to use either the model of Brailsford and Major or that of Cheng and Vachon; the former is simpler to apply arithmetically and since the calculation can only be approximate it is reasonable to use equation (6.8).

Unfortunately, real solids are rarely amenable to this simple treatment, for the so-called solid phase may well have an unknown microstructure and in addition may contain closed pores. In this case at least one experimental measurement of the thermal conductivity of a sample of known porosity is necessary to make predictions of the thermal conductivity of similar solids with differing porosities. This has been demonstrated by Austin (1939) who used Maxwell's equation with Russell's (1935) correction for radiation above 80% porosity. He interpreted with reasonable success the thermal conductivity of a series of diaspore bricks and of silica bricks as functions of porosity and temperature. In each case a roughly linear dependence of thermal conductivity as a function of porosity was found experimentally up to 50% porosity and the value extrapolated to zero porosity was used for the conductivity of the solid phase in Maxwell's model. The extrapolated value in diaspore bricks was lower than that of alumina and in the silica bricks was higher than that of quartz. Since the diaspore bricks were mainly large-grained corundum with small pores, the solid phase was assumed to be continuous and the thermal conductivity of the bricks over a full range of porosity at different temperatures was calculated. The theoretical values agreed reasonably well up to about 50% porosity and were slightly high at higher porosities. The silica bricks had a much larger pore structure than diaspore, with small grains of solid, so that a model with air as the continuous phase seemed more appropriate. The fit was not so good; the theoretical values were lower than the measured values at all porosities, although they tended to merge at higher porosities. It is probable that equation (6.8) would have been a better model over the full range in this case. However Austin's work has shown that if the thermal conductivity of a solid of one porosity is known, the equation of Maxwell (or Brailsford and Major) can be used to evaluate the effective conductivity of the solid phase and hence to estimate roughly the conductivities of similar solids of varying porosity. But it must be remembered that this is only a rough model, idealising the pores to a spherical shape. The shape and size of the pores could be of much greater importance than the total porosity—large disc-shaped pores, perpendicular to the heat flow would obviously have a much greater effect than a similar volume of air arranged as thin cylinders with their axes in the direction of heat flow.

Bricks are commonly used to provide high-temperature insulation, as furnace linings for example, where the porosity is responsible for the low conductivity. However, if the porosity is too high, radiation and convection will cause a rapid increase in apparent conductivity above 1000°C so that lightweight bricks will no longer be effective thermal insulators. At still higher temperatures radiation across small pores becomes an increasingly effective heat transfer mechanism, so that above 2000°C heat insulation can only be achieved by the use of dense materials. In this context pyrolytic graphite provides a very effective insulator to heat flow perpendicular to the plane of deposition. A comparatively new material in the field of high-temperature insulation is porous carbon. Measurements by Neuer and Wörner (1973) have shown carbon foam with 88% volume porosity to have a thermal conductivity of $\sim 0{\cdot}01$ W cm^{-1} K^{-1}, almost independent of temperature from 1400 to 2000 K, indicating the negligible effect of radiation over this range. At 2000 K it thus has a thermal conductivity equal to that of pyrolytic graphite in the *c*-direction. Carbon foam could therefore be a useful insulator at very high temperatures. More data are required to find its operating limit which will be the temperature at which either radiative transfer across the pores becomes significant or the structure itself becomes degraded.

The behaviour of rocks, which are porous solids, is of increasing importance in the field of geophysics. The problem is complicated because the fluid in the pores may well be a liquid rather than air. Since water, for example, has a thermal conductivity some 25 times that of air, the presence of water greatly increases the thermal conductivity. Moreover, if a liquid phase is present, the thermal conductivity of the rocks is likely to show an abrupt increase at temperatures low enough to freeze the saturating fluid. Woodside and Messmer (1961) and Brailsford and Major (1964) have tried to interpret measurements on dry and water-saturated porous sandstone using the Maxwell model. In both cases they found that, although the porosities were less than 60%, the experimental values were lower than those calculated with the solid phase assumed to be continuous. Brailsford and Major also found that for the sandstone saturated with water the experimental values were even lower than those calculated for the fluid phase assumed to be continuous, and were close to the minimum value for the mixture predicted by the series model. They interpreted this by postulating that not all the pores were connected; thus, although in the dry state all were filled with air, some were inaccessible to water so that only partial filling occurred. They replaced the porosity ϕ in equation (6.8) by an apparent porosity ϕ_{app}, where

$$\phi_{app} = \phi[\chi + (1-\chi)\phi]; \tag{6.13}$$

χ is a parameter which was adjusted to fit the measurements on dry sandstone. The results calculated for wet sandstone using the same value of χ fitted the experimental results quite closely. Sugawara and Yoshizawa (1962) used an empirical approach to the same measurements on sandstone and showed that the results could be fitted to the relationship:

$$\lambda_{mix} = (1-K)\lambda_{sol} + K\lambda_{fluid} ,$$

where

$$K = \frac{2^n[1-(1+\phi)^{-n}]}{2^n - 1} ; \tag{6.14}$$

λ_{sol} and λ_{fluid} are the conductivities of the solid and fluid respectively and n is an adjustable parameter. However, the physical significance of this parameter in terms of the structure of the solid is not known.

Increase in the conductivity of a porous solid is caused not only by the presence of water within its pores but also as a result of the transfer of the latent heat of evaporation by the diffusion of water vapour in the air in the pore space. This gives rise to an effective conductivity within the pore due to latent heat transfer, which increases rapidly as the temperature increases owing to an increase in the saturated vapour pressure. Krischer (1941) has shown that at about 340 K the rate of heat transfer is as great as if the pores were filled with water (see figure 6.8), so that at this temperature the thermal conductivity of damp pores should be independent of the extent to which the pores are filled with water. Above this temperature heat should be transferred at a higher rate in a moist pore containing air than completely filled with water. However, the overall effect of the transfer of latent heat on the thermal conductivity of a porous solid will be considerably less than the increase in the conductivity of the damp pore. It has also been found that the amount of moisture in a solid depends to some extent on whether it was brought into equilibrium with its surroundings from a dry or a wet state.

The presence of moisture leads to problems in the steady state measurements of thermal conductivity, for moisture is transferred from the warm to the cold region, condensation occurs, and water can drain out of the sample if heat flow is downward through the sample. This means that, if a guarded-hot-plate apparatus is used, the thermal conductivity of the upper specimen will be higher than that of the lower specimen. In addition, there can be an exchange of moisture between the samples and the surroundings. It is possible to maintain a uniform moisture distribution only when the temperature gradient across the sample is very small.

The formation of zones of varying moisture concentration which accompanies heat flow through a damp porous solid is aided by capillary forces. As the pore size is decreased the capillary force becomes larger than the viscous resistance to water flow, so that capillary attraction tends to draw water into the fine pores from the coarser ones. At low moisture content only the very small pores tend to hold water, the larger pores becoming wet as the moisture content is increased. Consequently in damp porous solids there may be some pores with dry walls, some completely filled with water, and some with damp walls. Krischer and Esdorn (1956) have based a model on such a structure and developed a

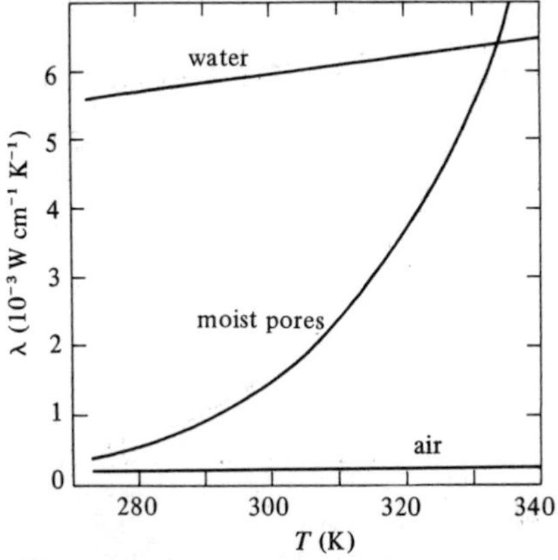

Figure 6.8. Thermal conductivity of moist pores compared with that of air and of water.

relationship for the thermal conductivity of a porous moist solid which includes empirical parameters to allow for the relative volumes of dry and damp pores.

The presence of moisture is of great importance in considering the heat transfer through building materials, for most of these, including bricks, masonry, and concrete, are porous and exposed to the elements. Use of the conductivity of the dry material in designing a comfortable human environment can lead to an appreciable underestimate of the heat losses. Consequently too small a heating plant may be provided, with the result that the temperature will be too low for comfort, or alternatively the fuel consumption will be higher than estimated. It is common practice in Great Britain to assume that external walls will have a 5% moisture content and inside walls one of 1% or 3% for brick or concrete respectively. If thermal conductivity data at the required moisture content are not available, they are predicted from an empirical relationship of Jakob (1949), the results of which are summarised in table 6.2.

The thermal insulation characteristics of a solid improve as the proportion of air in it is increased; as long as the pores within it are smaller in size than about 3 mm, convection of the gas is negligible and the greater the porosity the closer the thermal conductivity of the solid approaches that of still air. Consequently the materials with the lowest conductivities, as seen in figure 6.1, are all porous.

In order to improve the insulation of buildings, lightweight concretes are frequently used. These contain either lightweight aggregates or are made from a special lime, cement, and silica mixture aerated by the addition of a foaming agent which produces a concrete with a cellular structure, usually called an aerated or gas concrete. As the strength of the material diminishes with increase in voidage, there is in practice a limit to the porosity which can be tolerated, and it may be necessary to compromise between the thermal insulation required and the necessary strength. The Building Research Station has examined a large volume of data from many sources and found that for design purposes the thermal conductivity as a function of dry density can be considered to be

Table 6.2. Correction factors for the conductivity of moist building materials.

Water content (vol.%)	Multiplying factor to correct dry value
1	1·30
2·5	1·55
5	1·75
10	2·10
15	2·35
20	2·55
25	2·77

independent of the type of concrete; their recommended values (Arnold, 1970) are shown in figure 6.9. Although some experimental values deviate from this line by as much as 50%, there is no obvious correlation between the deviation and the type of concrete. Some lightweight aggregates contain very large voids and they should be avoided.

Other materials finding increasing use as thermal insulators in many applications, including building, are the cellular plastics. Since the solid plastic with its short range order has itself a very low conductivity, it is particularly suitable as a basic material. Plastics such as expanded polystyrene and foamed urea formaldehyde have room temperature conductivities between $3{\cdot}3 \times 10^{-4}$ and 4×10^{-4} W cm^{-1} K^{-1}. However, their upper operating temperature is limited and lies between 50 and 120°C, depending upon the particular polymer.

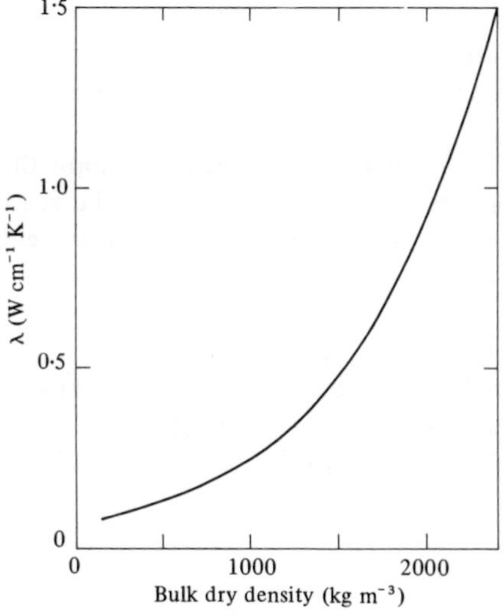

Figure 6.9. Thermal conductivity of concrete as a function of density.

6.3.3 Fibrous materials

It has been shown that thermal insulators are mainly porous and, if radiative effects are neglected, the thermal conductivity at high porosities tends towards that of the fluid within the pores. Improved insulation can be made by replacing the continuous solid structure by a fibrous mat consisting of fibres which may be randomly distributed in layers or in three dimensions. Heat flow through the solid takes place along a tortuous path through the fibres and is constricted by small regions of contact between the adjoining fibres, so that the effective conductivity of the solid is considerably less than that of the bulk material. Under these conditions the heat transported by conduction, convection, and radiation through the continuous paths can be more important than the heat flowing through the solid path. Such lightweight fibrous solids approximate to the series model with a thermal conductivity given by equation (6.5). If the conductivities of air and solid are represented by λ_1 and λ_2 respectively, and the volume fraction ϕ of air is large, then this equation approximates to

$$\lambda_{\mathrm{mix}} = \frac{\lambda_1}{\phi} , \tag{6.15}$$

where λ_1 must also include heat transfer by convection and radiation. To be slightly more realistic, we must make allowance for the heat flow through the solid path; a mathematical model for a fibrous insulation was proposed by Speil (1964):

$$\lambda_{\mathrm{mix}} = \frac{\lambda_{\mathrm{gas\,eff}} + \lambda_{\mathrm{r}} + \lambda_{\mathrm{conv}}}{\phi} + \lambda_{\mathrm{sol\,eff}} , \tag{6.16}$$

where $\lambda_{\mathrm{gas\,eff}}$, λ_{r}, and λ_{conv} are effective conductivities due to gas conduction, radiation, and convection respectively. $\lambda_{\mathrm{sol\,eff}}$ is the effective conductivity of the solid.

The thermal conductivity of a confined gas is dependent upon the gas pressure when the dimensions of the space occupied by the gas are comparable to the mean free path of the gas molecules. Consequently

$$\lambda_{\mathrm{gas\,eff}} = \lambda_{\mathrm{gas}} \frac{L}{\ell + L} , \tag{6.17}$$

where λ_{gas} is the conductivity of the free gas, L is an average pore dimension of the insulation, and ℓ is the mean free path. If ℓ_0 is the mean free path at unit pressure, then at pressure p, $\ell = \ell_0/p$, and equation (6.17) can be written

$$\lambda_{\mathrm{gas\,eff}} = \lambda_{\mathrm{gas}} \frac{1}{1 + \ell_0/Lp} . \tag{6.18}$$

The quantity $(1 + \ell_0/Lp)^{-1}$ is plotted in figure 6.10 as a function of p for various values of L. For a particular value of L it varies in magnitude from 0 to 1 over a range in pressure of approximately 100:1. In fibrous insulations the pore size is usually 10^{-5} cm or greater so that at normal pressure $\lambda_{\mathrm{gas\,eff}}$ is equal to the thermal conductivity of the free gas λ_{gas}.

The radiant energy transfer depends upon reflection and absorption; in the latter mode energy radiated from a warm surface is intercepted and then re-radiated at a different wavelength in all directions rather than directionally to the colder surface. Both mechanisms depend upon fibre diameter. Theoretically the problem is complex and many approaches to it have been made. In general it is found theoretically and experimentally that the magnitude of the radiative component varies directly with the cube of the mean absolute temperature T_{mean}, and with the diameter d of the fibre, and inversely with the volume fraction f occupied by the solid, that is inversely as the density of the insulation. The model proposed by Verschoor and Greebler (1952) yields

$$\lambda_r = \frac{2\cdot 15 \times 10^{-13} T_{\mathrm{mean}}^3 d}{f\xi^2}, \qquad (6.19)$$

where ξ is an opacity factor representing the absorption integrated over the spectral distribution of radiation. Other derivations vary slightly in the magnitude of the constant and in the part played by the absorption characteristics of the solid. As shown by Hager and Steere (1967), as a first approximation the fibres can be assumed to be black and ξ equal to unity. Although λ_r diminishes as the fibre diameter is decreased, at very small diameters, of the order of a few microns (that is of the same order as the dominant infrared wavelengths), scattering and absorption will tend to decrease, so that λ_r is likely to pass through a minimum and then increase at still smaller pore diameters.

Also convection is dependent upon the size of the pores within the insulation and decreases as the fibre density is increased. Empirical estimates of the contribution by convection in fibrous materials has been made by Allcut (1951) and by Verschoor and Greebler (1952).

Theoretical estimations of $\lambda_{\mathrm{sol\,eff}}$ have been made on the basis of a symmetrical lattice of uniform fibres. Hager and Steere assume there is no constriction at the contact between fibres and find that

$$\lambda_{\mathrm{sol\,eff}} = 4\lambda_{\mathrm{sol}} f^3, \qquad (6.20)$$

a relationship which tends to overestimate $\lambda_{\mathrm{sol\,eff}}$, but which allows a rough estimate to be made of its magnitude and is simpler to use than the method of Strong *et al.* (1960) which has also been found to give too large a value.

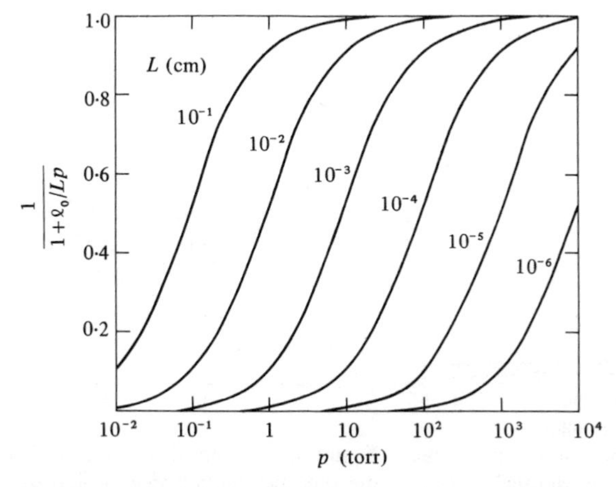

Figure 6.10. The quantity $(1-\ell_0/Lp)^{-1}$ as a function of pressure.

The relative importance of the various mechanisms is shown in figure 6.11 (after Pratt, 1969) as a function of the density of the fibrous insulation. It can be seen that at 338 K air conduction is the predominant mechanism up to a density of 140 kg m^{-3}, while radiation and convection are only appreciable below densities of about 20 kg m^{-3}. Although air conduction predominates in most fibrous materials around room temperature, this curve is not unique. The magnitude of the various components depends on the fibre diameter and to some extent on the properties of the fibre material, and these can vary in materials of equal density, but the figure does give a general indication of behaviour. At still higher densities solid conduction increases and the total conductivity goes through a minimum. At higher temperatures the radiation contribution with its T^3 dependence becomes increasingly important, and the density for minimum conductivity of a particular material increases, because a greater density is necessary to reduce the radiation to a point where the sum of the contributions has a minimum value. The general behaviour is shown in figure 6.12, where it can also be seen that the effective conductivity of the fibrous materials increases with increasing temperature. It is obvious that no generalisation can be made concerning the density of fibrous material which will give the highest insulation; the material must be picked in light of the operating conditions.

It should be noted that the insulation properties of fibrous materials are anisotropic. This is particularly marked when the fibres are randomly oriented in layers, when the conductivity for heat flow perpendicular to the layers is much lower than when the heat flow is along the layers.

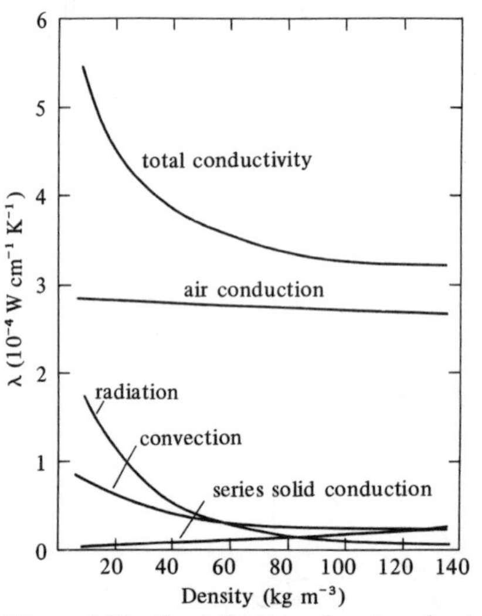

Figure 6.11. Contribution of various heat flow mechanisms to the thermal conductivity of fibrous insulation.

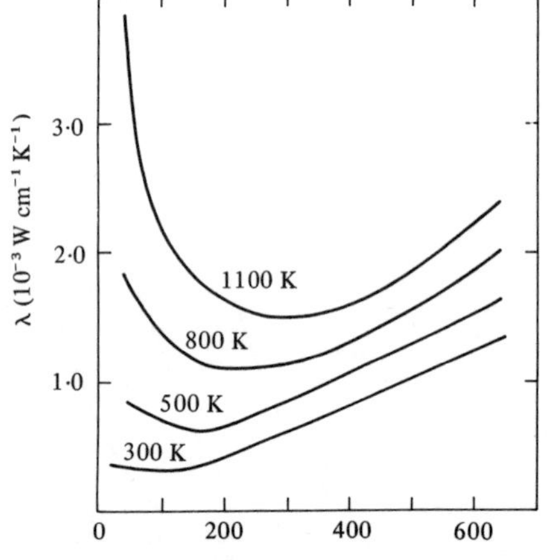

Figure 6.12. Typical relationship between the thermal conductivity of fibrous insulation and its density.

The insulation performance of fibrous materials is obviously improved by evacuation, as the air conduction and convection components are then reduced to zero. This is frequently done in cryogenic engineering, but it is not a feasible proposition in many applications. An alternative to pressure reduction is the replacement of air by a gas of greater molecular weight, to make the mean free path of the molecules comparable with the pore size and thus lower the effective gas conductivity as shown by equation (6.18).

6.3.4 Superinsulations

The previous section on fibrous materials has shown how the effective conductivity of a solid can be reduced by constricting the heat flow paths through fine fibre contacts with the result that gaseous conduction predominates. Recent developments have produced solids where both the solid and gaseous conductions are reduced. This has been achieved by the manufacture of highly porous solids with a predominance of pores of the order of 10^{-5} cm or less in diameter, with the result that the effective conductivity of the solid is less than that of still air. A typical curve for the conductivity of Microtherm, one of these so-called microporous insulators, is given as a function of gas pressure in figure 6.13. The fact that the curve is still rising at 10^3 torr indicates, by comparison with figure 6.10, the presence of pores of the order of 10^{-6} cm. The variation covers some four orders of magnitude of pressure owing to the presence of a range of pore sizes. At normal pressures the thermal conductivity is rather more than half that of still air. The conductivity at the lowest pressure is that due to heat transfer by solid conduction and by radiation. These high-performance insulators consist basically of microporous silica with the addition of other ceramic oxides opaque to infrared radiation

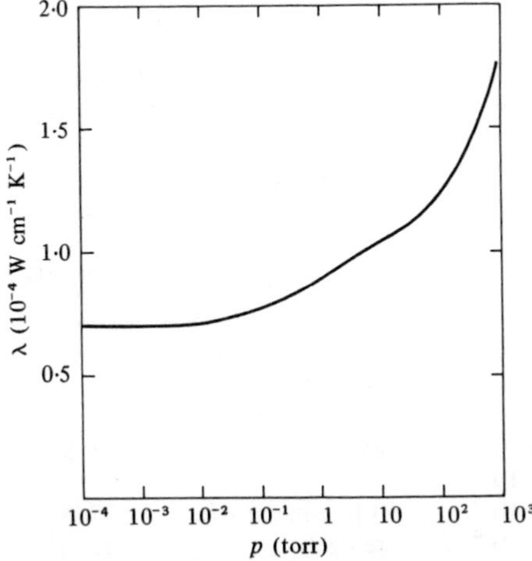

Figure 6.13. Thermal conductivity of Microtherm as a function of gas pressure.

and with large scattering and absorption cross sections, so that radiative heat transport is not excessive. They can be produced in various forms of solid or blanket, the precise conductivity depending upon the form. A typical conductivity is plotted against mean temperature in figure 6.1. The operation of these solids is usually restricted to a maximum hot face temperature of around 1000°C owing to degradation of their properties.

Superinsulating composite systems consisting of very thin highly-reflecting metal foils, separated by fine-powder or fibrous insulation, have been developed. The smallest possible fibres or fine, preferably microporous, particles are used, so that there are an enormous number of point contact resistances in the solid path. In addition, as in Microtherm, the chosen solid must be opaque to infrared radiation with large scattering and absorption cross sections.

These insulations were developed primarily for aerospace projects and they are all expensive to manufacture. The multilayer types tend to be used mainly in cryogenic applications and are evacuated so that there is no gaseous conduction. Operation above room temperature tends to damage the reflecting foils; these are either aluminium or plastic films coated with evaporated silver or gold so as to reduce their emissivity and thereby increase the impedance to radiative transfer. Similar systems are in use above room temperature though they are not so effective. Suitable refractory metals provide the radiation shields, but their emissivity is not as low as that of aluminium or the silver and gold coatings. Moreover, high-temperature operation can damage their reflecting nature and impair the properties of the powder or fibre separating the shields.

6.3.5 Powders

Thermal conductivity is very dependent upon physical texture, as illustrated by table 6.3, which compares the bulk conductivity of various solids with that in some porous form. The low values obtained are due not only to the presence of air but also to the constriction to heat flow in the solid path at contacts between grains or fibres. Both glass silk and carbon black have conductivities less than that of still air owing to

Table 6.3. Comparative thermal conductivities of various solids in bulk and porous form.

Solid	λ at 300 K (W cm^{-1} K^{-1})		Porous form	Effective porosity (%)
	bulk	porous		
Steel	0·58	$4{\cdot}1 \times 10^{-3}$	$\frac{1}{8}$ in dia. spheres	41·3
Lead	0·33	$4{\cdot}2 \times 10^{-3}$	$\frac{1}{16}$ in dia. spheres	42
Alumina	0·36	$4{\cdot}0 \times 10^{-3}$	166 μm particles	42
Aluminium	2·31	$7{\cdot}2 \times 10^{-4}$	wool	98·5
Glass	0·0094	$3{\cdot}6 \times 10^{-4}$	silk	93
Carbon	1·0	$2{\cdot}5 \times 10^{-4}$	0·01 μm particles	97
Silica	0·014	$2{\cdot}2 \times 10^{-4}$	aerogel	95

the presence of a large volume of pores smaller than the mean free path of air molecules. Powders are frequently used as thermal insulators, not only in the superinsulation multilayer structures, but also in more general applications, around furnace tubes for example. As low a conductivity as possible is required, and in general the smaller the particle size the lower the conductivity.

Various attempts have been made to correlate experimental results with the models discussed in section 6.3.1, particularly that of Maxwell (with gas phase continuous) either for spheres or modified to spheroids. None give good agreement; the theoretical predictions in general are too low at room temperature and the experimental conductivities all increase more rapidly with temperature than the model predicts owing to neglect of radiative transfer through the air spaces.

Laubitz (1959) interpreted experimental measurements on powders of MgO, Al_2O_3, and ZrO_2 of uniform particle size, in terms of a two-phase conductivity (using a somewhat similar model to that of Maxwell) together with a radiative term. However, only if the conductive term was arbitrarily doubled could he fit the results, and the model failed to agree with measurements on graded powders. As Laubitz himself pointed out, although the mathematics is exact, the model is so artificial that it radically departs from real powders.

In an attempt to bring the mathematical model closer to the real powder, Godbee and Ziegler (1966) derived an expression for the conductivity of a two-phase system which included a shape factor based on a logarithmic normal distribution of particles, which accounted for particle shape and the packing factor. Using shape factors obtained experimentally from the particle-size distribution curves and including radiative effects, they were able to explain successfully results on MgO, Al_2O_3, and ZrO_2 powders.

A rather different approach has been made by Luikov *et al.* (1968), based on an analogue of series and parallel resistors, which allows for such things as constriction resistance, microroughness and oxide film at the powder contacts, gas conduction through the pores and the spaces between the projections of roughness on the contacting surfaces, and radiative transfer. The theoretical expression is very complex but reasonable agreement has been obtained with published data over a wide range of temperature and pressure on many different powders and granular materials.

6.3.6 Composite materials

Over the years composite materials such as crystalline solids in a glassy matrix, cermets, and fibre-reinforced solids have all found increasing applications.

The cermets are heterogeneous combinations of metals or alloys with one or more ceramic phases which constitute from 15% to 85% of the solid by volume, and in which there is little solubility between the various phases. Very fine fibres of a material have great tensile strength, approaching the theoretical limit and when incorporated in a material tend to give the composite a greater strength than the continuous matrix.

The thermal conductivity of a composite is either higher or lower than that of the major phase, depending on whether the thermal conductivity of the additional phase is greater or smaller than that of the continuous matrix. Since the shape presented to heat flow by inclusions, such as fibres, depends upon the direction within the composite, its thermal conductivity is likely to be anisotropic and could be markedly so.

In order to predict the thermal conductivity, one must consider the microstructure of the material and choose an appropriate model from those summarised by Reynolds and Hough (1957), or from equations given by Behrens (1968) which also allow for the shape of the dispersed constituent. An alternative type of model for a statistically isotropic heterogeneous medium has been proposed by Lakkad *et al.* (1972) in which a two-phase system is reduced to an equivalent polycrystalline aggregate. It is possible that under some conditions this model could predict thermal conductivity within narrower limits than the others. None of the theoretical relationships, however, make allowance for any interactions at the surfaces between the various phases; these could produce high interfacial resistances. Measurements on ceramic–metal laminates by Francis and Tinklepaugh (1960) established that, owing to interfacial resistances, the thermal conductivity perpendicular to the laminae was 10% lower, and parallel to the laminae 12% higher, than calculated. The magnitude of such effects would depend not only on the magnitude of the interfacial resistance but also on the structure of the composite.

Accurate knowledge of the thermal conductivity of composites can only be obtained experimentally, but for many purposes sufficiently accurate predictions can be made, as long as the chosen mathematical model is a reasonable approximation to the structure of the composite.

References

Allcut, E. A., 1951, *Proceedings of a General Discussion on Heat Transfer* (Institute of Mechanical Engineers, London).
Arnold, P. J., 1970, Building Research Station Current Paper 1/70.
Austin, J. B., 1939, ASTM STP, **39**, 3.
Behrens, E., 1968, *J. Composite Materials,* **2**, 2.
Brailsford, A. D., Major, K. G., 1964, *Brit. J. Appl. Phys.,* **15**, 313.
Cheng, S. C., Vachon, R. I., 1969, *Int. J. Heat Mass Transfer,* **12**, 249.
Cosgrove, G. J., McHugh, J. P., Tiller, W. A., 1961, *J. Appl. Phys.,* **32**, 621.
Francis, R. K., Tinklepaugh, J. R., 1960, *J. Am. Ceram. Soc.,* **43**, 560.
Godbee, H. W., Ziegler, W. T., 1966, *J. Appl. Phys.,* **37**, 56.
Hager, N. E., Steere, R. C., 1967, *J. Appl. Phys.,* **38**, 4663.
Jakob, M., 1949, *Heat Transfer* (Chapman and Hall, London), Vol.1.
Kelly, B. T., 1969, in *The Chemistry and Physics of Carbon* (Dekker, New York), Vol.5.

Kingery, W. D., 1959, *J. Am. Ceram. Soc.,* **42**, 617.
Krischer, O., 1941, *Wärme- u. Kältech.,* **43**, 2.
Krischer, O., Esdorn, H., 1956, *Forsch. Gebiete Ingenieurw.,* **22**, 1.
Lakkad, S. C., Miatt, B. B., Parson, B., 1972, *J. Phys. D, Appl. Phys.,* **5**, 1304.
Laubitz, M. J., 1959, *Canad. J. Phys.,* **37**, 798.
Loeb, A. L., 1954, *J. Am. Ceram. Soc.,* **37**, 96.
Luikov, A. V., Shashkov, A. G., Vasiliev, L. L., Fraiman, Yu. E., 1968, *Int J. Heat Mass Transfer,* **11**, 117.
Maxwell, J. C., 1904, *A Treatise on Electricity and Magnetism* (Clarendon Press, Oxford).
Moore, J. P., McElroy, D. L., 1971, *J. Am. Ceram. Soc.,* **54**, 40.
Neuer, G., Wörner, B., 1973, *High Temperatures - High Pressures,* **5**, 279.
Pratt, A. W., 1969, in *Thermal Conductivity,* Ed. R. P. Tye (Academic Press, New York), Vol.1, p.316.
Reynolds, J. A., Hough, J. M., 1957, *Proc. Phys. Soc. B,* **70**, 769.
Russell, H. W., 1935, *J. Am. Ceram. Soc.,* **18**, 1.
Simpson, A., Stuckes, A. D., 1971, *J. Phys. C, Solid State Phys.,* **4**, 1710.
Speil, S., 1964, *Applied Materials Res.,* **3**, 238.
Strong, H. M., Bundy, F. P., Bovenkerk, H. P., 1960, *J. Appl. Phys.,* **31**, 39.
Sugawara, A., Yoshizawa, Y., 1961, *Austral. J. Phys.,* **14**, 469.
Sugawara, A., Yoshizawa, Y., 1962, *J. Appl. Phys.,* **33**, 3135.
Touloukian, Y. S. (ed.), 1970, *Thermophysical Properties of Matter, The TPRC Data Series* (IFI/Plenum Press, New York), Vols.1 and 2.
Ure, R. W., 1962, *J. Appl. Phys.,* **33**, 2290.
Verschoor, J. D., Greebler, P., 1952, *Trans. Am. Soc. Mech. Engrs.,* **74**, 961.
Woodside, W., Messmer, J. H., 1961, *J. Appl. Phys.,* **32**, 1699.

Appendix

Conversion factors for thermal conductivity

	$W\ m^{-1}\ K^{-1}$	$W\ cm^{-1}\ K^{-1}$
1 $W\ m^{-1}\ K^{-1}$	1	10^{-2}
1 $W\ cm^{-1}\ K^{-1}$	10^{2}	1
1 cal $cm^{-1}\ s^{-1}\ K^{-1}$	418·68	4·1868
1 kcal $m^{-1}\ h^{-1}\ K^{-1}$	1·163	$1·163 \times 10^{-2}$
1 Btu $ft^{-1}\ h^{-1}$ deg F^{-1}	1·731	$1·731 \times 10^{-2}$
1 Btu in $h^{-1}\ ft^{-2}$ deg F^{-1}	0·1442	$1·442 \times 10^{-3}$

Author index

Italicized numbers refer to pages, at the end of each chapter, on which references are listed in full.

Subject index

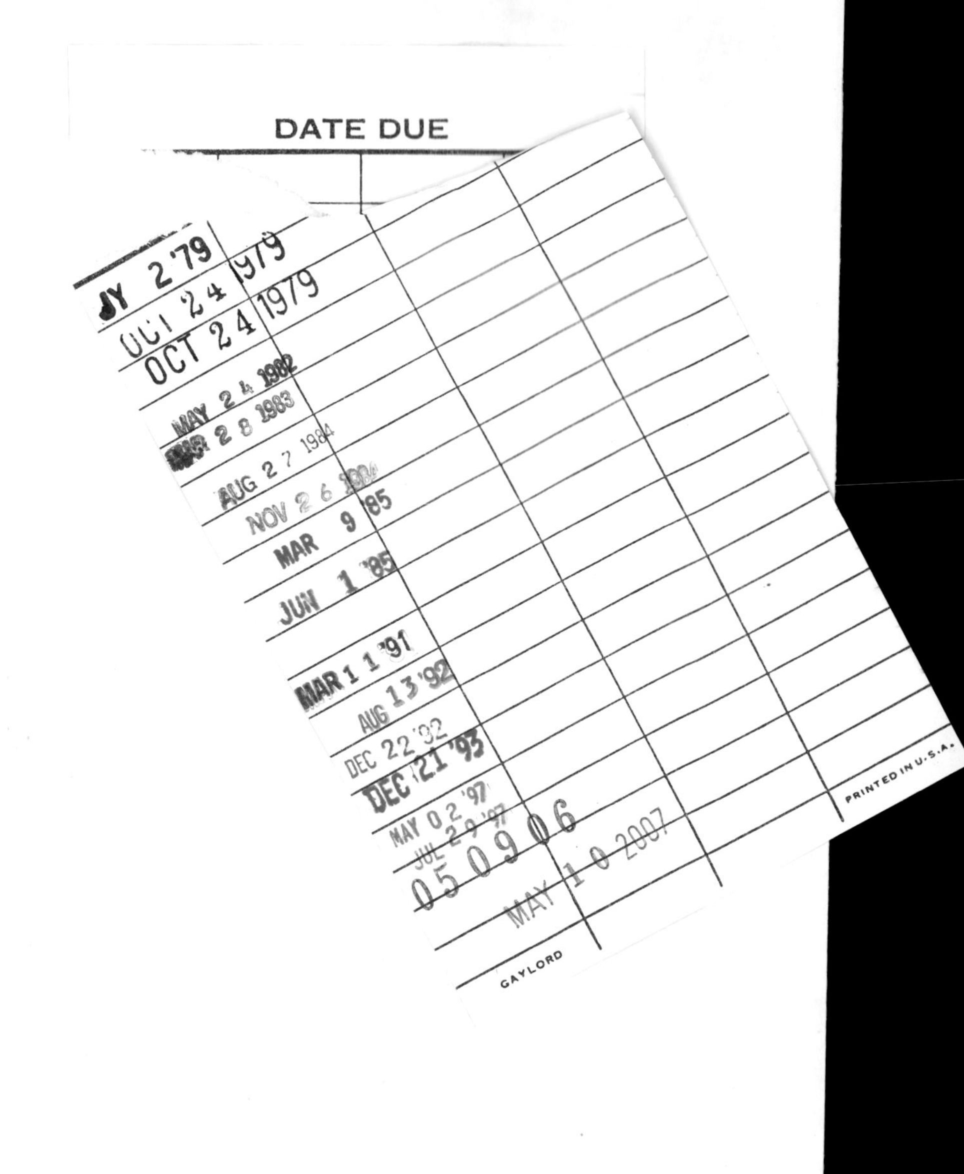